Materials, Manufacturing Processes and Devices

Materials, Manufacturing Processes and Devices

Editor

Peng Cao

University of Auckland, New Zealand

World Scientific

NEW JERSEY • LONDON • SINGAPORE • BEIJING • SHANGHAI • TAIPEI • CHENNAI

Published by

World Scientific Publishing Co. Pte. Ltd.
5 Toh Tuck Link, Singapore 596224
USA office: 27 Warren Street, Suite 401-402, Hackensack, NJ 07601
UK office: 57 Shelton Street, Covent Garden, London WC2H 9HE

British Library Cataloguing-in-Publication Data
A catalogue record for this book is available from the British Library.

MATERIALS, MANUFACTURING PROCESSES AND DEVICES

ISBN 978-981-98-1842-6 (hardcover)
ISBN 978-981-98-1843-3 (ebook for institutions)
ISBN 978-981-98-1844-0 (ebook for individuals)

For any available supplementary material, please visit
https://www.worldscientific.com/worldscibooks/10.1142/14456#t=suppl

Desk Editor: Muhammad Ihsan Putra

Typeset by Stallion Press
Email: enquiries@stallionpress.com

https://doi.org/10.1142/9789819818433_fmatter

Preface

This monograph compiles a selection of 26 research papers that exemplify advancements in materials science, nanotechnology, and energy-related applications. 19 papers were selected from a special issue of the *International Journal of Modern Physics Letters B*, featuring works presented at the 2nd International Symposium on Materials, Manufacturing Processes, and Devices (MMPD-2019), held at the University of Auckland, New Zealand, from 15th to 17th February 2019.

MMPD-2019 was organized by the University of Auckland in collaboration with Tsinghua University, Southwest Jiao Tong University, Changzhou Institute of Technology, and Jiangsu University of Science and Technology, with sponsorship from the Society of Materials New Zealand and Auckland Uniservices. The symposium provided a dynamic forum for presenting groundbreaking research in nanostructured materials, advanced coatings, and device technologies for applications in energy, catalysis, and sensing.

The other 7 papers were selected from recent issues of *International Journal of Modern Physics Letters B*. They include insightful studies on nanocomposites for energy storage, thermoelectric materials, photonic bandgap structures, and novel manufacturing techniques. Key highlights include enhanced thermal conductivity in nanofluids, advanced supercapacitor designs, graphene-based materials, and novel approaches to lithium-ion battery anodes.

Together, these works reflect the interdisciplinary nature of modern materials science, spanning the synthesis and characterization of nanostructures to the development of next-generation energy devices and sustainable manufacturing methods. We hope this collection will serve as an invaluable resource for researchers, educators, and industry professionals, driving further innovation in these vital fields.

We extend our heartfelt thanks to the authors, reviewers, and collaborators whose efforts made this publication possible, as well as to the institutions and sponsors that supported MMPD-2019.

Peng Cao
*The University of Auckland,
New Zealand*
18 November 2024

https://doi.org/10.1142/9789819818433_fmatter

Contents

https://doi.org/10.1142/9789819818433_0001

Chapter 1

Enhanced Thermal Conductivity of Nanofluids Made of Metal Oxide Nanostructures Synthesized by Arc Discharge Method#

F. Fang*,‡, J. Futter*, E. Hutchinson*, J. Leveneur*,† and J. Kennedy*,†

*National Isotope Centre, GNS Science,
P. O. Box 31-312, Lower Hutt 5040, New Zealand

†The MacDiarmid Institute for Advanced Materials and Nanotechnology,
Victoria University of Wellington,
P. O. Box 600, Wellington 6140, New Zealand

‡v.fang@gns.cri.nz

We have modified the mineral oil used in transformers by dispersing 1-wt.% metal oxide nanostructures (commercially available TiO_2, ZnO and Fe_3O_4 spherical structures and ZnO rod-shaped structures synthesized by arc discharge) into the oil through ball milling without surfactant. A good dispersion that lasted for at least 24 h was obtained for all nanofluids, however sedimentation was discovered by 72 h after ball milling. All nanofluids with different nanostructures exhibited enhanced thermal conductivity compared with the raw transformer oil. The nanofluid with ZnO nanoparticles showed better thermal conductivity than the nanofluids with TiO_2 and Fe_3O_4 nanoparticles. The nanofluid with elongated ZnO nanoparticles (nanorods) synthesized by arc discharge showed the best thermal conduction among all the nanofluids studied in this work over the whole measurement period. The enhanced thermal conductivity of the nanofluid with elongated nanostructure is considered to be due to the rod-shaped nanostructure creating heat flow paths with lower thermal resistance. The arc discharge provides a cost-effective and scalable method to fabricate metal oxide nanostructures for potential nanofluid applications.

Keywords: Metal oxide; arc discharge; nanofluid.

‡Corresponding author.

#To cite this article, please refer to its earlier version published in the *International Journal of Modern Physics B*, Volume 34, 2040001 (2020), DOI: 10.1142/S0217979220400019.

1. Introduction

Heat transfer is a critical element in the design of industrial processes and consumer products. The efficiency of heat transfer can have a significant impact on industrial, commercial or residential carbon footprints. For example, cooling energy costs can account for more than 50% of a commercial building's electricity consumption because of the heat produced by the high-process-density circuit boards required for enhanced functionality in modern telecommunications.[1] Efficient heat transfer processes can be used to reduce electricity consumption through waste heat recovery. Nanofluids, nanoparticle suspensions in conventional fluids, have emerged as a new generation of thermal fluids with superior heat transfer capabilities. The literature has suggested that incorporation of nanoparticles in fluids tends to enhance heat transfer performance, with significant effects on the heat transfer coefficient, thermal conductivity and the viscosity, which affects the required fluid pumping power.[2,3]

Metal oxide nanoparticles are preferred over metals for nanofluids because they are more chemically stable. The lower density and electrical conductivity of metal oxide nanoparticles indicate that they are less prone to particle settling and agglomeration.[4–8] From a commercial production perspective, metal oxide nanoparticles have great potential for use in nanofluids as they can be more easily produced than other nanoparticles and are relatively cheap.[9,10] Nanofluids made with metal oxide nanoparticles, e.g., alumina (Al_2O_3),[11,12] silicon dioxide (SiO_2),[13] copper oxide (CuO)[14] and titanium oxide (TiO_2),[15] have received particular attention.

The objective of this work was to study the effect of metal oxide nanospherical and elongated nanoparticle structures on the thermal conductivity of nanofluids. The nanofluids were designed as dispersions of 1% by mass of different metal oxide nanostructures, including zinc oxide (ZnO), TiO_2 and iron oxide (Fe_3O_4) spherical nanoparticles. For the fourth nanofluid, arc discharge was used to synthesize ZnO elongated nanoparticles (nanorods) in an effort to enhance thermal conduction. Our approach will involve the use of elongated nanospherical structures to create paths with a lower thermal resistance for heat flow, which leads to the enhanced thermal conduction in nanofluids.

2. Experimental

Nanofluids in this work were prepared using commercially available ZnO, TiO_2 and Fe_3O_4 nanoparticles together with ZnO rod-shaped

Table 1. Metal oxide nanostructures used in this work for preparing nanofluids.

Material	Purity	Average size	Shape	Supplier
TiO_2 (Rutile)	99.9%	∼ 50 nm (Ref. 16)	Spherical	Inframat
ZnO	99.7%	∼ 30 nm (Ref. 17)	Spherical	Advanced
Fe_3O_4	99.9%	20–50 nm (Ref. 18)	Spherical	Materials LCC[19]
ZnO nanorods	99.9%	Length: 0.5–1 μm; diameter: 40–400 nm	Elongated rod-shaped	GNS Science[20]

nanostructures synthesized by an arc discharge method at GNS Science. Details of these nanostructures are provided in Table 1. Detailed information on the arc discharge synthesis of ZnO nanorods can be found in Ref. 20. The nanostructure concentration was kept at 1 wt.% for all nanofluid samples. The host fluid used was Shell Diala transformer oil which is a highly refined mineral oil. The nanofluids were prepared by dispersing nanostructures into transformer oil by ball milling for 24 h without using any surfactant.[21]

The thermal conductivity of the nanofluids was compared using a modified hot-wire method constructed in-house. Two identical 40-mL test tubes were used, one containing pure transformer oil and the other containing nanofluids. Each tube was heated with a copper wire coil driven by a constant-voltage source with the current running through both coils in series ensuring equal energy input. The temperature of each coil was monitored by a "K"-type miniature thermocouple attached to the coil with thermally conductive epoxy. A LABview App written in-house was used to graph and log results to CSV output files for later data presentation.

3. Results and Discussion

The ZnO nanorods (Fig. 1) show lengths of 0.1–5 μm and diameters between 40 nm and 400 nm with high purity.[20,22] Nanostructures tend to agglomerate due to a combination of attractive and repulsive forces.[23] Ball milling and stirring were explored in this work to disperse nanostructures in the transformer oil for better performance. Figure 2 clearly demonstrates that both stirring and ball milling can achieve well-dispersed nanofluids. However, the stability of the nanofluids varies. Particle settling occurred within 12 h at rest after stirring the TiO_2 nanofluid for 5 h. The TiO_2 nanofluid prepared by ball milling still showed a homogeneous dispersion after resting for 24 h. Particle settling in this sample was observed after resting for 72 h.

Fig. 1. SEM image of ZnO nanorods synthesized by the arc discharge method.

Fig. 2. (Color online) Stability of nanofluids with TiO_2 nanoparticles dispersed by stirring and ball milling: (a) immediately after stirring for 5 h, (b) rest for 12 h after stirring for 5 h, (c) rest for 24 h after ball milling for 24 h and (d) rest for 72 h after ball milling for 24 h.

Colloidal stability of nanofluids plays a vital role in the suitability of nanofluids in real-world applications. The Brownian velocity of the nanostructures should be appreciably the stability of nanofluids, surfactants are often used to modify the surface of nanoparticles.[7,24] The surfactants or coupling agents have attractive interactions with both the nanoparticles and the fluid and locate at the interface between the two phases, improving contact and reducing interfacial energy. The surfactants can also provide electrostatic repulsion between the particles and/or steric barriers to prevent agglomeration.[24] To improve the stability of the nanofluids studied in this work, it is recommended that hydrophobic groups or silane coupling agents be used to modify the metal oxide nanoparticle surfaces.

During the thermal conductivity measurements, a thin metallic wire coil served as a heat source. The metallic wire was immersed in the test fluid and is heated by passing a current through it. The temperature rise of the metallic wire was influenced by the thermal conductivity of the fluid. The higher the thermal conductivity of the fluid, the lower is the temperature rise of the metallic wire.[8]

Figure 3 shows the change in temperature of the metallic wire with heating time for the nanofluids prepared with commercially available spherical nanoparticles. Temperature fluctuation was discovered on pure transformer oil after it was heated and the temperature increase became smooth after 10 min. It was considered due to the delayed temperature circulation in the pure transformer oil. All the nanofluids in this work exhibited enhanced thermal conductivity compared with the raw transformer oil. The temperature of pure transformer oil increased from room-temperature (22°C) to 38°C in 3000 s. In that time, the nanofluids containing 1-wt.% ZnO nanoparticles and TiO_2 nanoparticles only increased to 25°C and 27°C, respectively. The nanofluid with 1-wt.% Fe_3O_4 nanoparticles showed

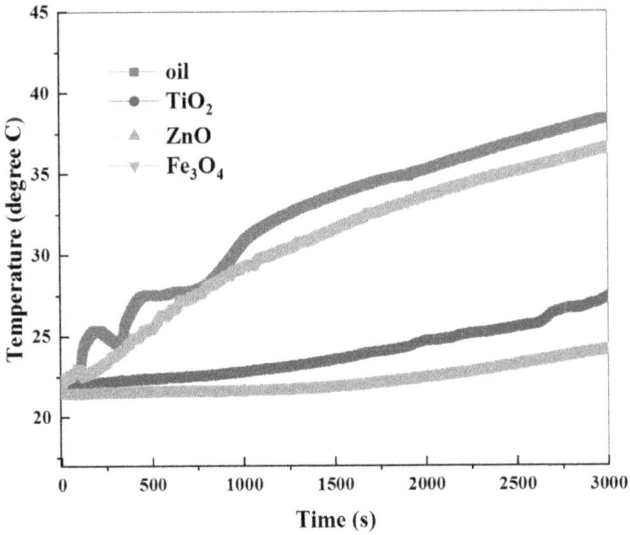

Fig. 3. (Color online) Temperature variationsous with time for nanofluids with and without nanoparticles.

a similar temperature increase as the raw transformer oil. Of the nanofluids prepared with the commercial spherical nanoparticles, the ZnO nanoparticle nanofluid showed the best thermal conductivity performance which may be due to the best Brownian motion it has; this is because Brownian motion of nanoparticles induces microconvection, resulting in superior thermal performance of nanofluids compared to the base fluid. Other factors, such as particle size, nanolaying and the effect of nanoparticles clustering,[25] may also have contributed to the better thermal conductivity performance of ZnO nanoparticle nanofluid than other metal oxide nanoparticle nanofluids studied in this work.[26]

A nanofluid with 1-wt.% ZnO nanorods synthesized by arc discharge method was also fabricated. Figure 4 shows the change in temperature of the metallic wire with heating time when immersed in the two different ZnO nanofluids. The metallic wire has a lower temperature in the nanofluid with ZnO nanorods during the whole measurement period. The temperatures of the metallic wire in the two nanofluids were very similar after heating, and the temperature difference started to be obvious only after heating for 600 s. After heating for 3000 s, the metallic wire temperatures for the nanofluids with ZnO nanoparticles and ZnO nanorods were 24.11°C and 23.58°C, respectively.

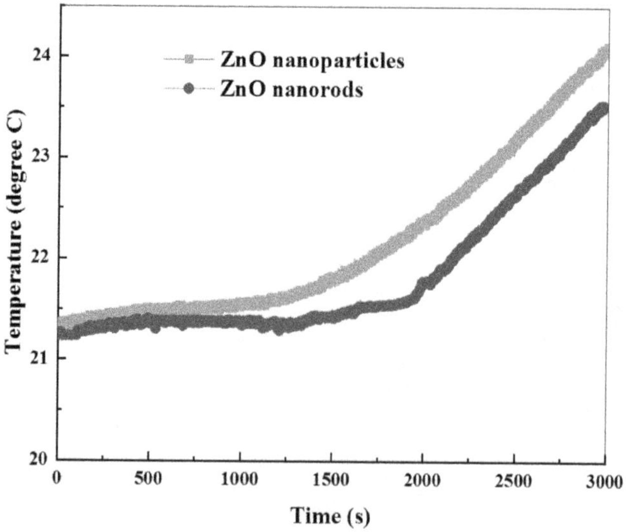

Fig. 4. (Color online) Temperature variationsous with time for nanofluids with ZnO nanoparticles and ZnO nanorods.

The results indicate that the nanofluid with arc discharge-synthesized ZnO nanorods shows enhanced thermal conduction compared with the nanofluid with the commercial ZnO spherical nanoparticles. A possible explanation is that the rod-shaped nanostructures create a path with lower thermal resistance for heat flow, which leads to the enhanced thermal conduction in nanofluids.[27] Several other studies have also shown that nanofluids with elongated nanoparticles have a higher thermal conductivity than nanofluids with spherical-shaped structures. Ferroulliat *et al.* studied the effect of nanoparticle shape on thermal properties and thermal performance of porous SiO_2–water and ZnO–water nanofluids.[28] They found that nanofluids with rod-shaped particles had higher thermal conductivities than those with polygonal-shaped particles. Colangelo *et al.* prepared water- and diathermic oil-based nanofluids with different shaped particles (CuO, Al_2O_3 spherical and ZnO elongated) and reported that the nanofluids with elongated-shaped ZnO nanoparticles had the highest thermal conductivity enhancement at the same particle volume concentration.[29] It can be recommended that the elongated structures are more efficient in enhancing the thermal conductivity than spherical structures of the same volume fraction. This agrees well with the results discovered both experimentally and theoretically in Ref. 30 that geometry is one of the main factors controlling thermal conductivity enhancement in nanofluids.

The improved heat transfer performance of the metal oxide nanofluids can be attributed to various molecular-level mechanisms, including particle size, micro-convection induced by Brownian motion, interfacial liquid layering and particle clustering. The influence of such molecular-level mechanisms on thermal conductivity of nanofluids depends on various physical factors, such as geometry, agglomeration state, nanoparticle–liquid interaction, temperature, viscosity, etc. Detailed investigation of the changes occurring in the nanofluids is needed to understand the temperature dependency of nanofluids.

4. Conclusion

In summary, this work compared the thermal conductivities of nanofluids prepared with metal oxide nanostructures both in spherical shape and elongated rod shape. The incorporation of these nanoparticles enhanced the thermal conductivity of transformer oil.

The nanofluid with ZnO nanoparticles showed better thermal conductivity than the nanofluids with TiO_2 and Fe_3O_4 nanoparticles. This was considered due to better Brownian motion and other proper factors, such as nanolaying and the effect of nanoparticles clustering, that ZnO nanoparticles brought into the nanofluid. The nanofluid with elongated ZnO nanoparticles (nanorods) synthesized by arc discharge showed better thermal conduction than the nanofluid with spherical ZnO and other metal oxide nanoparticles over the whole measurement period. The enhanced thermal conductivity is considered due to the rod-shaped geometry creating heat flow paths with lower thermal resistance. The arc discharge provides a cost-effective and scalable method to fabricate metal oxide nanostructures for potential nanofluid applications.

Acknowledgments

This work was performed under a research contract (C05X1713) from the Ministry of Business, Innovation and Employment (MBIE).

References

1. J. Yang, K.-T. Chan and X. Wu, in *Proc. Sixth Int. Conf. Enhanced Building Operations (ICEBO2006)* (2006).
2. S. K. Das *et al.*, *J. Heat Transf.* **125**, 567 (2003).
3. Y. Li *et al.*, *Powder Technol.* **196**, 89 (2009).
4. K. Kwak and C. Kim, *Korea-Aust. Rheol. J.* **17**, 35 (2005).
5. L. Fedele *et al.*, *Nanoscale Res. Lett.* **6**, 300 (2011).
6. P. Keblinski, R. Prasher and J. Eapen, *J. Nanopart. Res.* **10**, 1089 (2008).
7. W. Yu *et al.*, *Heat Transf. Eng.* **29**, 432 (2008).
8. K. S. Suganthi and K. S. Raja, *Renew. Sustain. Energy Rev.* **76**, 226 (2017).
9. D. Cabaleiro *et al.*, *Nanoscale Res. Lett.* **8**, 286 (2013).
10. L. Fedele, L. Colla and S. Bobbo, *Int. J. Refrig.* **35**, 1359 (2012).
11. A. K. Starace *et al.*, *J. Appl. Phys.* **110**, 124323 (2011).
12. S. Q. Zhou and R. Ni, *Appl. Phys. Lett.* **92**, 093123 (2008).
13. D. Shin and D. Banerjee, *J. Heat Transf.* **135**, 032801 (2013).
14. H. O'Hanley *et al.*, *Adv. Mech. Eng.* **2012**, 181079 (2012).
15. H. M. Nieh *et al.*, *Int. J. Therm. Sci.* **77**, 252 (2014).
16. Inframat Advanced Materials LLC, Titanium dioxide (Titania, TiO_2) nano powder, rutile (2019), http://www.advancedmaterials.us/22N-0814R.htm.
17. Inframat Advanced Materials LLC, Zinc oxide (ZnO) nano powder ($\sim$ 30 nm) (2019), http://www.advancedmaterials.us/30N-0801.htm.
18. Inframat Advanced Materials LLC, Iron(II, III) oxide (Fe_3O_4) nano powder (2019), http://www.advancedmaterials.us/26FO23-08N1.htm.
19. Inframat Advanced Materials LLC, Inframat: About us (2019), http://www.advancedmaterials.us.
20. F. Fang *et al.*, *Nanotechnology* **20**, 245502 (2009).

21. V. Segal *et al.*, *J. Magn. Magn. Mater.* **201**, 70 (1999).
22. J. Kennedy *et al.*, *Diam. Relat. Mater.* **71**, 79 (2017).
23. G. J. M. Koper, *An Introduction to Interfacial Engineering* (Delft Academic Press, The Netherlands, 2007).
24. S. Manikandan *et al.*, *Braz. J. Chem. Eng.* **29**, 699 (2012).
25. Y. Xuan, Q. Li and W. Hu, *AIChE J.* **49**, 1038 (2003).
26. S. U. S. Choi *et al.*, in *Proc. DOE BES 20th Symp. Energy Engineering Sciences* (2002).
27. P. Keblinski *et al.*, *Int. J. Heat Mass Transf.* **45**, 855 (2002).
28. S. Ferrouillat *et al.*, *Appl. Therm. Eng.* **51**, 839 (2013).
29. G. Colangelo *et al.*, *Appl. Energy* **97**, 828 (2012).
30. E. V. Timofeeva *et al.*, *Phys. Rev. E* **76**, 061203 (2007).

https://doi.org/10.1142/9789819818433_0002

Chapter 2

Synthesis of NiO/TiO_2 Sub-microrods as Catalyst for Methanol Oxidation Reaction[#]

Xiaoyu Yue, Yixuan Wang, Ting Zhang and Wei Gao*

Department of Chemical and Materials Engineering,
The University of Auckland, Auckland 1142, New Zealand
**w.gao@auckland.ac.nz*

NiO/TiO_2 sub-microrods have been synthesized via a simple two-step route. First, the precursors were prepared by a facile ethylene glycol-mediated method; then, NiO/TiO_2 sub-microrods were obtained by a limited-oxygen atmosphere deriving from the decomposition of urea at 450°C for 3 h in air. The samples were characterized by X-ray diffraction (XRD), scanning electron microscopy (SEM), energy-dispersive spectroscopy (EDS) and transmission electron microscopy (TEM). The obtained NiO/TiO_2 sub-microrods exhibit cubic NiO structure with high crystallinity and anatase TiO_2. Both SEM and TEM show the typical sub-microrods with lengths of $\sim$ 380 nm and diameters of $\sim$ 180 nm. The uniform sub-microrods have great electrocatalytic performance for methanol oxidation reaction in alkaline solution. This material may have potential applications in direct methanol fuel cells.

Keywords: NiO/TiO_2 sub-microrods; ethylene glycol-mediated method; methanol oxidation.

1. Introduction

As one of the alternative energy sources, direct methanol fuel cell is a good option to replace fossil fuels and reduce environmental pollution. Pt-based catalysts have been known as the most effective electrocatalysts for methanol oxidation. However, the commercialization is hindered by several drawbacks such as poor kinetic reaction and high cost of Pt.[1]

*Corresponding author.

[#]To cite this article, please refer to its earlier version published in the *International Journal of Modern Physics B*, Volume 34, 2040002 (2020), DOI: 10.1142/S0217979220400020.

Recently, transition metal oxides including NiO,[2] CoO (Ref. 3) and TiO_2 (Ref. 1) have been studied as the methanol oxidation catalysts or catalyst supports. These metal oxides exhibit good stability and anti-poisoning ability in acid or alkaline solutions. Ordered NiO/TiO_2 nanotube arrays have been studied as catalyst support.[4]

In this work, NiO/TiO_2 sub-microrods were prepared through ethylene glycol-mediated method, which exhibited good methanol electro-oxidation activity. These hybrid sub-microrods could be potential candidates as electrocatalyst for methanol oxidation.

2. Experimental

The samples were prepared by a modified process[2] as shown in Fig. 1. First, 50-mg PVP and 357-mg nickel(II) chloride hexahydrate were dissolved in 35 mL of ethylene glycol at room-temperature to form a clear solution. Then 2-mL tetrabutyl titanate was added dropwise to the mixture under vigorous stirring. Ten minutes later, 155 mg of sodium hydroxide was added to the above mixture under vigorous stirring followed by continuous stirring for 48 h. After being washed by ethanol for three times, the prepared green precursors were dried at 60°C. Finally, yellow samples were obtained by annealing at 450°C for 3 h with urea.

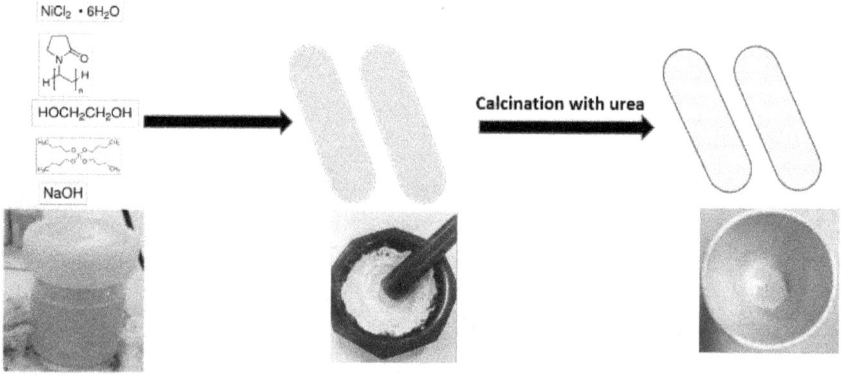

Fig. 1. (Color online) Schematic procedure for fabricating NiO/TiO_2 sub-microrods.

3. Results and Discussion

A typical X-ray diffraction (XRD) pattern of the synthesized composite is shown in Fig. 2(a). The diffraction peaks located at 37°, 43°, 63°, 75° and 79° correspond to (111), (200), (220), (311) and (222) planes, respectively, which can be indexed to the cubic NiO with FCC crystalline structure (JCPDS card no. 89-7130).[5] Other peaks at 25°, 48° and 54° are characterized as anatase TiO_2 (101), (200) and (105) planes (JCPDS card no. 71-1169).[6] No impurity peaks were observed, indicating that the final material was a combination of TiO_2 and NiO with high purity. Figure 2(b) is the transmission electron microscopy (TEM) micrograph of NiO/TiO_2 sub-microrods, and it shows that the products are regular rods with length and diameter of ca. 350 nm and 180 nm, respectively. The structure

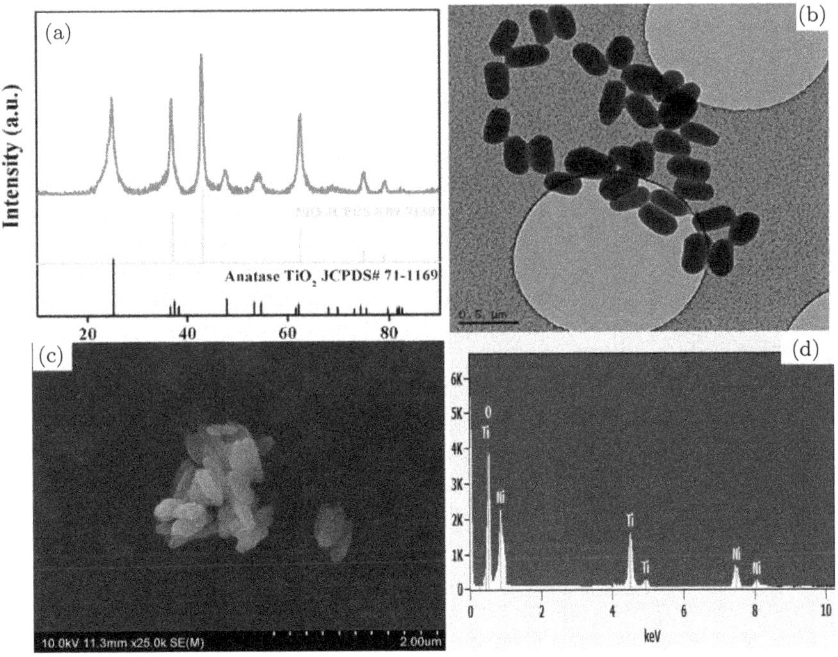

Fig. 2. (Color online) (a) XRD pattern, (b) TEM micrograph, (c) SEM image and (d) EDS pattern of NiO/TiO_2 sub-microrods.

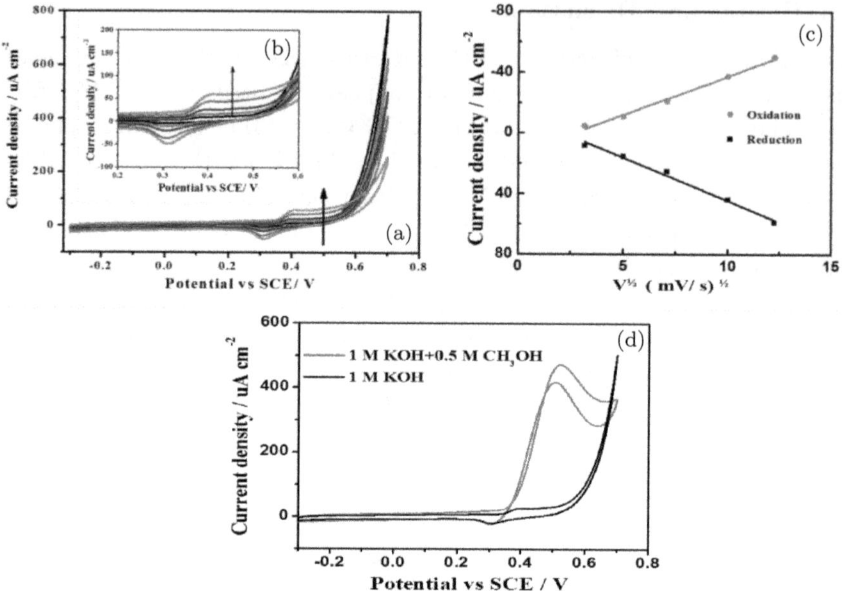

Fig. 3. (Color online) (a) and (b) The CVs of NiO/TiO_2 catalyst in 1.0M KOH concentration at different scan rates of 10, 25, 50, 100 and 150 $mV{\cdot}s^{-1}$; (c) the relationship between redox peak current and square root of scan rate; and (d) the CVs of NiO/TiO_2 catalyst in 1.0M KOH with and without 0.5M methanol at a potential sweep rate of 50 $mV{\cdot}s^{-1}$.

of as-prepared composite was further observed by scanning electron microscopy (SEM) and the observation is presented in Fig. 2(c), which is in agreement with the TEM result. The energy-dispersive spectroscopy (EDS) pattern [Fig. 2(d)] of the surface layer confirms the existence of the elements Ni, Ti and O.

Figures 3(a) and 3(b) show the CV plots of NiO/TiO_2 sub-microrod electrodes in 1.0M KOH at different scan rates from 10 $mV \cdot s^{-1}$ to 150 $mV \cdot s^{-1}$. The catalyst exhibits obvious pair of redox peaks between 300 mV and 500 mV, which are assigned to the Ni(II)/Ni(III) redox group. It was found that both anodic and cathodic current densities increased linearly with the square root of the scan rate as shown in Fig. 3(c), indicating a diffusion-controlled process. Similar behavior has been reported for NiO-based materials.[4] The electro-oxidation mechanism of methanol over NiO materials is as follows:

$$NiO + OH^- \rightarrow NiOOH + e^-, \quad (1)$$

$$NiOOH + Methanol \rightarrow Ni(OH)_2 + Oxidation\ products. \quad (2)$$

To evaluate the catalytic performance, CV measurements of methanol oxidation were conducted in the presence and absence of 0.5M methanol in 1.0M KOH [Fig. 3(d)]. Without methanol, the catalyst displayed the Ni(II)/Ni(III) redox couple which indicated the formation of electroactive species on the surface of NiO/TiO$_2$ sub-microrods, such as the high-valance NiOOH. However, after the addition of methanol, the redox couple disappeared, and a typical electrocatalytic response of methanol oxidation reaction occurred. The CV curve shows the methanol oxidation current peak (I_f) at ca. 500 μA/cm^2 in the forward scan and an oxidation peak (I_b) at ca. 400 μA/cm^2 in the backward scan, corresponding to the removal of the residual carbonaceous species formed in the forward scan. The higher I_f/I_b value indicates higher tolerance to intermediate carbon species, which means methanol can be oxidized to carbon dioxide much more efficiently.[7] The ratio for the NiO/TiO$_2$ sub-microrods is calculated to be 1.2. The oxidation of methanol appeared with the onset potential of 0.36 V. The catalytic ability and tolerance to CO are much better than the NiO–CoO electrode.[3]

4. Conclusion

In summary, we developed an effective method for the preparation of uniform NiO/TiO$_2$ sub-microrods. The synthesis process includes a facile ethylene glycol-mediated method and post-treatment in a limited-oxygen atmosphere deriving from the decomposition of urea at 450°C for 3 h in air. The catalyst was investigated for its electrocatalytic activity against methanol oxidation reaction in basic media. The sub-microrods show good electrocatalytic activity with a low onset potential at ca. 0.36 V for methanol electro-oxidation and good tolerance towards CO in comparison with other catalysts. The enhancement could be attributed to the high crystallinity of NiO and the internal interaction between TiO$_2$ and NiO. The above facts indicate that the as-prepared NiO/TiO$_2$ hybrid structures can be a potential candidate as an electrocatalyst for methanol oxidation reaction.

Acknowledgments

The authors would like to acknowledge the assistance of the Staff in the Department of Chemical and Materials Engineering, The University of Auckland, New Zealand. Xiaoyu Yue would also like to thank the China Scholarship Council (CSC) for the Ph.D. Scholarship.

References

1. X. Yue *et al.*, *Electrochim. Acta* **668**, 174 (2015).
2. J. Wang *et al.*, *J. Mater. Chem. A* **20**, 9948 (2017).
3. F. Hong *et al.*, *J. Cluster Sci.* **29**, 670 (2018).
4. H. He *et al.*, *Phys. Status Solidi A* **212**, 2090 (2015).
5. G. B. Anandha *et al.*, *Opt. Magn. Prop.* **25**, 41 (2013).
6. L. Shan *et al.*, *Thin Solid Films* **516**, 3059 (2008).
7. Y. J. Li *et al.*, *Carbon* **48**, 1128 (2010).

https://doi.org/10.1142/9789819818433_0003

Chapter 3

Oxygen Vacancies-Contained SnO_2 Nanotubes with Enhanced Ethanol Gas Sensing Properties[#]

Yifan Chen[*,†], Xiuling Ma[†], Chen Li[†], Qiuyu Wu[†], Yongbo Wang[†], Menjun Huang[‡], Bitao Liu[†,§,‖] and Lu Li[†,¶,‖]

[*] *Faculty of Materials and Energy, Southwest University, Chongqing 400700, P. R. China*

[†] *Research Institute for New Materials Technology, Chongqing University of Arts and Sciences Chongqing 402160, P. R. China*

[‡] *Chongqing Institutes of Higher Education Key Forensic Science Laboratory, Southwest University of Political Science and Law, Chongqing 401120, P. R. China*

[§] *liubitao007@163.com*

[¶] *lli@cqwu.edu.cn*

SnO_2 porous nanotubes containing oxygen vacancies were prepared by electron spinning and H plasma treatment. The morphology and crystal structure of the samples were analyzed by scanning electron microscopy (SEM), transmission electron microscopy (TEM) and X-ray diffraction (XRD). The ethanol-sensing properties of the SnO_2 sensor were tested. The results show that the samples treated with H plasma for 20 min have the best performance. Its working temperature is 230°C lower than 290°C of the original sample, with a sensitivity of 17 at 100 ppm, which is seven times higher than the original sample. It also shows good selectivity to some common interfering gases. This enhancement can be ascribed to the introduced oxygen vacancy. This work provides an efficient way to design high-performance gas sensor materials.

Keywords: Oxygen vacancies; porous nanotubes; ethanol.

[‖] Corresponding authors.

[#] To cite this article, please refer to its earlier version published in the *International Journal of Modern Physics B*, Volume 34, 2040003 (2020), DOI: 10.1142/S0217979220400032.

1. Introduction

The declining energy resources and environmental degradation are crucial problems worldwide. Detection of gases which threaten the environment and human health has attracted growing attention.[1–3] Metal–oxide–semiconductor-based gas sensors have been extensively researched for various applications because of their excellent sensing performance, even in harsh environments.[4,5] Among the rest, tin dioxide (SnO_2) has been proved to be prospective for gas sensing by virtue of its outstanding photoelectrical properties, superior thermal stability (melting point is 1127°C), chemical inertness, low cost and nontoxicity.[6–8] However, its working temperature usually reached 300°C, and the sensitivity is still lower for some industrial applications. Thus, it is a challenge to improve the sensor properties of SnO_2-based materials.[5–15]

Recently, SnO_2 with different nanostructures, especially the 1D nanostructured metal oxides, has been reported for detecting various types of gases.[9] It holds great promise in diverse applications because of its large surface-to-volume ratio, facile electron transport, relatively good structural stability and potential to build higher-dimensional architectures. It is well-known that gas sensing is based on the surface interaction process between the oxide surface and target gas molecules.[10–15] Thus, the surface structure design is the key point for the gas sensor. For example, oxygen vacancy can provide electro-optical interaction by providing reactive surface sites, and it would enhance its sensing efficiency.[15,16] Herein, we report an electrospinning method to prepare SnO_2 nanotubes and introduce oxygen vacancy by hydrogen (H) plasma treatment. The result indicates that the oxygen vacancies-contained SnO_2 has a better gas sensitivity for ethanol.

2. Experimental

The SnO_2 is synthesized by the method detailed in our previous work.[4] The obtained SnO_2 is further treated by hydrogen plasma with a power of 80 W, a flow rate of 20 mL/min and a treatment time of 10 min (HPT10), 20 min (HPT20) or 30 min (HPT30), respectively.

The powder X-ray diffraction (XRD) patterns were obtained on a Dandong TD-3500 X-ray diffractometer using Cu-Kα radiation. The morphology was analyzed using scanning electron microscopy (SEM; SU-8010; Hitachi) and transmission electron microscopy (HR-TEM; JEM-2100). X-ray photoelectron spectroscopy (XPS) measurements were recorded on

ESCA Lab MKII. The gas sensing performances were measured in an intelligent gas sensor analysis system (CGS-8; Beijing Elite Tech Co.).

3. Results and Discussion

The XRD patterns of the original sample and samples treated with hydrogen plasma are shown in Fig. 1. All diffraction peaks could be indexed to a tetragonal rutile structure of SnO_2 (JCPDS card no. 99-0024). As can be seen, the three characteristic peaks at 26°, 34° and 52° correspond to the crystal planes (110), (101) and (211), respectively. The absence of miscellaneous peaks indicates that the original sample is pure SnO_2. In addition, it can also be seen that the samples treated with H plasma contain the same characteristic peaks as the original SnO_2, indicating that the H plasma does not change the phase. Thus, to investigate the influence of H plasma treatment on the surface lattices, TEM and XPS were further utilized.

Figure 2 shows the SEM and TEM images of original SnO_2 and HPT20 samples. The nanotube structure composed of nanoparticles can be seen in Figs. 2(a) and 2(b). The diameters of the nanotubes are between 150 nm and 170 nm, and the size of the nanoparticles is in the range of 10–20 nm.

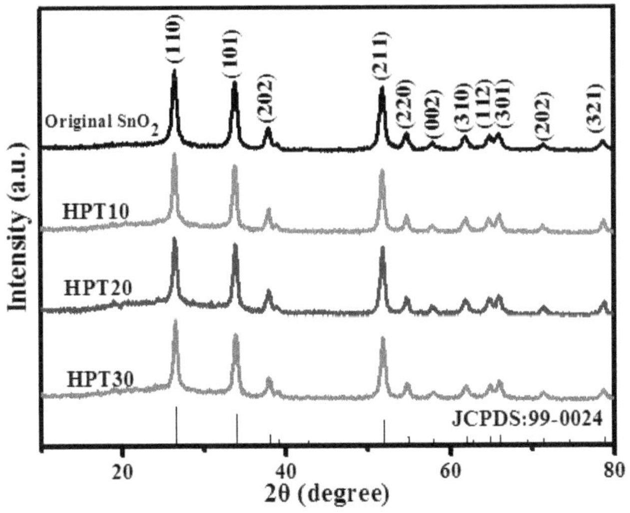

Fig. 1. (Color online) XRD patterns of original SnO_2 sample and SnO_2 samples treated with hydrogen plasma for different treatment times.

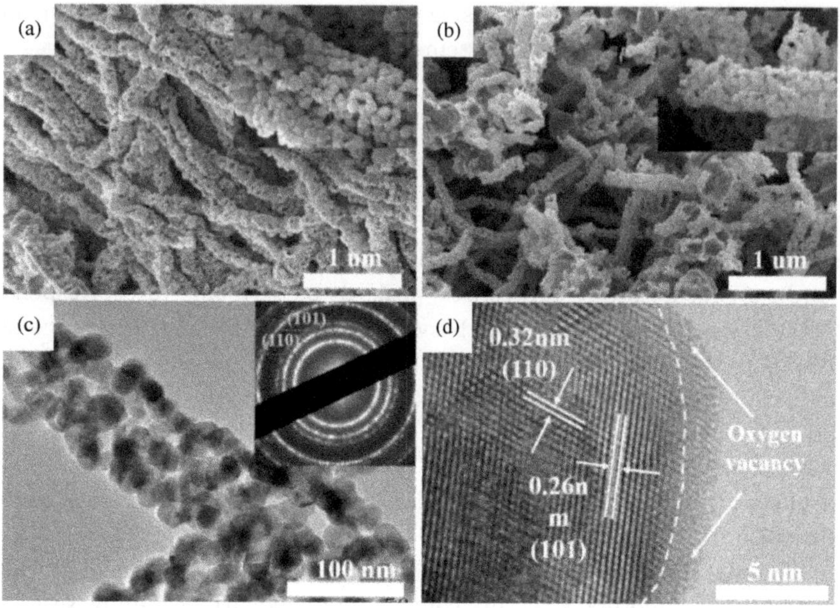

Fig. 2. (Color online) (a) SEM image of pure SnO_2; (b) SEM image of HPT20; (c) TEM image of HPT20 and (d) HR-TEM image of HPT20 (the inset scale bar is 200 nm).

Additionally, the H plasma-treated sample ruptures and breaks spatially, while the original sample remains in good shape and with certain length.

In order to find any surface structure changes, TEM was used to characterize the microstructure. Figures 2(c) and 2(d) show that the morphology of HPT20 nanotubes was the same as in Figs. 2(a) and 2(b). In addition, the hollow structure of these nanotubes also can be seen in Fig. 2(c). The HR-TEM image and SAED patterns of the samples in Fig. 2(d) show lattice fringes clearly, and the lattice spacings were 0.26 nm and 0.32 nm, corresponding to (101) and (110) planes of SnO_2, respectively. Furthermore, it can be seen that the surface lattice is disordered, indicating that the treated sample contains many surface defects. Obviously, these surface defects should come from the H plasma treatment, and would benefit the gas sensing performance.[11,14,16]

To demonstrate the presence of oxygen vacancy in plasma treatment, XPS measurements for the HPT20 sample were carried out and the results are shown in Fig. 3. The survey spectra of HPT20 and original SnO_2 samples both contain the peaks of C, O and Sn elements and show no significant

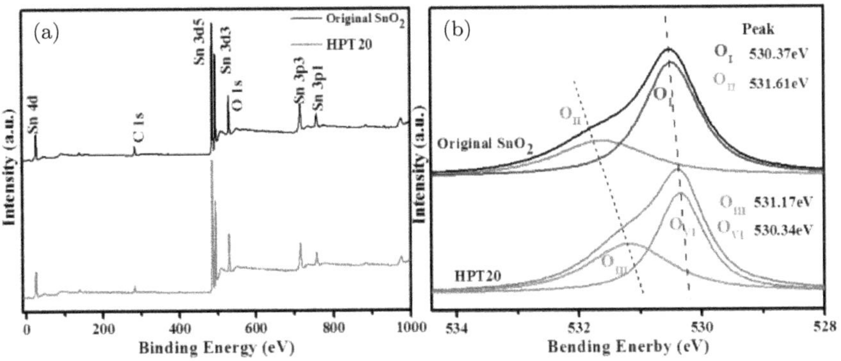

Fig. 3. (Color online) (a) XPS survey spectra of HPT20 and original SnO_2 samples and (b) O-1*s* spectra of HPT20 and original SnO_2 samples.

difference. Figure 3(b) shows a broader O-1*s* peak of original SnO_2 which is composed of O_I peak centered at 531.61 eV and O_{II} peak centered at 530.37 eV. The 530.37-eV peak is associated with the Sn–O–Sn lattice oxygen atoms in the rutile SnO_2 structure.[17] In HPT20, the position of this peak changes by only 0.03 eV, indicating that HPT20 is still SnO_2. In contrast, the position of the O_{II} peak was significantly shifted after H plasma treatment. The peak value migration from 531.61 eV to 531.17 eV of the combined energy cultivated land was related to O^{x-} ions (O^- and O^{2-} ions) adsorbed by oxygen vacancy in the anoxic zone of SnO_2 matrix.[17]

In order to determine an optimum operating temperature, the response of the sensor to 20 ppm of ethanol was examined. As shown in Fig. 4(a), the response of HPT20 increases with increasing operating temperature and attains a maximum of 7.92 at 230°C, while the maximum response of original SnO_2 sample was 2.19 at 290°C. The response of HPT20 is improved by 3.5 times, and the operating temperature is reduced by 60°C. It is known that oxygen molecules adsorb on the surface of SnO_2 nanotubes and form oxygen ions (O^-, O^{2-}) by capturing electrons from the conduction band, leading to a high-resistance state of SnO_2 nanofibers. Thus, the nanotube structure provides a high surface-to-volume ratio and large surface accessibility for the interaction between SnO_2 and ethanol, thus favoring the gas response by enhancing the Knudsen diffusion coefficient.[19] In addition, H plasma treatment can introduce oxygen vacancy,[11] thus making the sample absorb more oxygen molecules in the air to form oxygen species (O^- and O^{2-}), and improving its gas sensitivities.

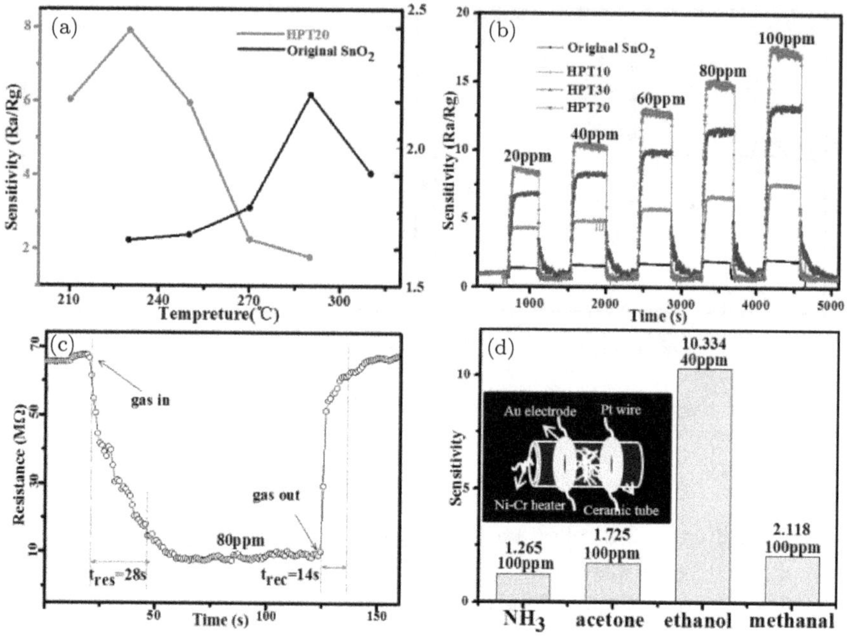

Fig. 4. (Color online) (a) Sensitivities at different temperatures, (b) sensitivities for different concentrations of ethanol, (c) dynamic response/recovery curves and (d) selectivity to different gas.

However, the long- and short-time treatments both would reduce the sensors' properties. Figure 4(b) shows the response/recovery properties at different processing times for 20–100 ppm of ethanol at the optimum operating temperature. The HTP20 has a better response of 17 at 100 ppm, which is eight times that of the original SnO_2. For HPT10 and HPT30 samples at 100 ppm, the responses are only 7.2 and 12.5, both lower than that of HPT20. This can be ascribed to less oxygen vacancies or heavy lattice disorder. Moreover, high response did not extend the response/recovery time of HPT20, but the response/recovery was more rapid.

Figure 4(c) shows an enlarged response/recovery loop at 80 ppm. The response time is 28 s and the recovery time is only 14 s, slightly below the average level of semiconductor oxide sensors.[14] The resistance level of the sensor after the response is basically the same as that before the response, showing that HPT20 has good stability and repeatability. At the same time, the characteristic of resistance variation of n-type semiconductor is presented. The selectivity test results for the common interfering gas concentration are shown in Fig. 4(d). At 40 ppm, the response to ethanol reached 10.334, while at 100 ppm, the responses to NH_3, acetone and formaldehyde were 1.265, 1.725 and 2.118, respectively.

4. Conclusions

SnO_2 porous nanotubes containing oxygen vacancies were prepared by electron spinning and H plasma treatment. The results showed that the H plasma treatment can decrease the working temperature to 230°C, lower than 290°C of the original SnO_2. The sensitivity is seven times higher than SnO_2. The response time is 28 s and the recovery time is only 14 s, slightly below the average level of semiconductor sensors. These samples also show good selectivity to some common interfering gases. The improvement can be ascribed to the introduced oxygen vacancy.

Acknowledgments

This work is supported by the Open Project of Chongqing Institutes of Higher Education Key Forensic Science Laboratory (XKZDSYS2019-Z2) and Chongqing Natural Science Foundation (cstc2017jcyjAX0259).

References

1. B. T. Liu *et al.*, *Electrochim. Acta* **292**, 407 (2018); C. R. Wu *et al.*, *Appl. Surf. Sci.* **441**, 1024 (2018).
2. M. W. G. Hoffmann *et al.*, *Adv. Mater.* **26**, 8017 (2014).
3. B. T. Liu *et al.*, *Nanoscale Res. Lett.* **11**, 391 (2016).
4. J. Cheng *et al.*, *Ceram. Int.* **43**, 301 (2017).
5. S. Singkammo *et al.*, *ACS Appl. Mater. Interfaces* **7**, 3077 (2015).
6. A. Das *et al.*, *J. Mater. Chem. C* **2**, 164 (2014).
7. J. H. Lee *et al.*, *ACS Appl. Mater. Interfaces* **7**, 3101 (2015).
8. Z. Y. Zhang *et al.*, *Nanoscale* **7**, 10078 (2015).
9. Y. Chen *et al.*, *J. Alloy. Compd.* **651**, 114 (2015).
10. Y. Wen *et al.*, *Nanoscale* **5**, 9739 (2013).
11. H. S. Shim *et al.*, *Sol. Energy Mater. Sol. Cells* **93**, 2062 (2009).
12. L. Liu *et al.*, *Mater. Lett.* **63**, 1975 (2009).
13. S. Piperno *et al.*, *J. Appl. Phys.* **101**, 124504 (2007).
14. F. Trani *et al.*, *Microelectron. J.* **40**, 236 (2009).
15. G. Pacchioni, *ChemPhysChem* **4**, 1041 (2003).
16. B. T. Liu *et al.*, *J. Power Sources* **429**, 46 (2019).
17. J. Zhan *et al.*, *Adv. Mater.* **28**, 795 (2016).

https://doi.org/10.1142/9789819818433_0004

Chapter 4

Dried Plum-Like ZnO Assemblies Consisted of ZnO Nanoparticles Synthesized by Ultrasonic Spray Pyrolysis#

Congzhi Zhang*, Tao Han†,‡, Wei Wang* and Jin Zhang†

*Chongqing BOE Display Lighting Co., Ltd., Chongqing 400714, P. R. China

†Research Institute for New Materials Technology, Chongqing University of Arts and Sciences, Chongqing 402160, P. R. China

‡danbaiht@126.com

Dried plum-like ZnO assemblies consisting ZnO nanoparticles were synthesized by an ultrasonic spray pyrolysis method (USP). ZnO assemblies were characterized by X-ray diffraction (XRD), field-emission scanning electron microscopy (FE-SEM), energy-dispersive spectroscopy (EDS) and field-emission transmission electron microscopy (TEM). The results show the size of ZnO assemblies is in the range of 300–870 nm, and that of ZnO nanoparticles is from 33 nm to 39 nm. The microstructure and size of ZnO assemblies were successfully controlled by the concentration of the precursor solution.

Keywords: ZnO assemblies; nanoparticles; ultrasonic spray pyrolysis method.

1. Introduction

In recent years, ZnO attracted extensive attention because it is a wide-bandgap semiconductor material. Due to the direct bandgap of 3.37 eV and large excitation binding energy of 60 meV, ZnO had various applications in optical and electronic fields such as in light-emitting diodes (LEDs), varistors, gas sensors, photocatalysts, transparent electronics and solar cells.[1–5]

‡Corresponding author.

#To cite this article, please refer to its earlier version published in the *International Journal of Modern Physics B*, Volume 34, 2040005 (2020), DOI: 10.1142/S0217979220400056.

A number of methods for synthesizing ZnO particles were investigated, such as the sol–gel method,[6] hydrothermal method,[7] co-precipitation method,[8] polyol process,[9] chemical vapor deposition[10] and ultrasonic spray pyrolysis (USP).[2] Among these processes, USP had the advantages of nontoxic nature, low cost and simple procedures, not only for the scientific experiment but also for the industrial production.

Up to now, USP has been used to produce metals and metallic oxides,[11] with the powder size in the range from microns to nanometers.[12] Milošević, Jordović and Uskoković prepared ZnO nanoparticles with smooth surfaces by USP at 900°C from zinc nitrate solution and ball ZnO particles with the size from $0.5\,\mu$m to $1.1\,\mu$m.[13] At various temperatures, Ebin *et al.* reported that ZnO nanoparticles with porous structure could be produced by USP. Flores–Carrasco *et al.* obtained ZnO nanoparticles in the range of 248 nm to 470 nm.[14] In this paper, dried plum-like ZnO assemblies consisted of ZnO nanoparticles were obtained by the USP method. The effects of concentrations on the size of ZnO assemblies are studied.

2. Experimental

2.1. *Material synthesis*

Zinc acetate dihydrate [$Zn(CH_3COO)_2{\cdot}2H_2O$] was purchased from Cheng Du Ke Long Chemical Reagent Co. All reagents were used as received without further purification. The dried plum-like ZnO assemblies consisting of ZnO nanoparticles were prepared in a furnace with two temperature zones. Different concentrations of Zn acetate dihydrate were used as the precursor solutions.

The USP system is composed of an ultrasonic atomizer, peristaltic pump, air compressor, a furnace with two temperature zones and a powder collector. ZnO assemblies were obtained through the following steps: (1) the precursor solution was atomized by the ultrasonic atomizer with a resonant frequency of 1.7 MHz. (2) The droplets are transported into the furnace with two temperature zones. The temperatures were set at 500°C and 700°C with an air speed of 6 SLPM of carrier gas. (3) The droplets were evaporated, pyrolyzed and then react in the high-temperature furnace. Finally, the obtained assemblies are carried into the powder collector by the carrier air.

2.2. *Characterization methods*

X-ray powder diffraction (XRD) tests were carried out on a diffractometer (TD-3500; Dandong, China) using monochromatized Cu-Kα radiation with an accelerating voltage of 30 kV and an applied current of 20 mA. Energy-dispersive X-ray spectrometry (EDS) maps were obtained to analyze the compositions of ZnO powders. Surface morphology and particle size were obtained by an field-emission scanning electron microscope (FE-SEM; JSM-7800F; JEOL) with an accelerating voltage of 5 kV and a field-emission transmission electron microscope (TEM; HT7700; Hitachi) with an accelerating voltage of 120 kV.

3. Results and Discussion

The crystalline structure and phases of samples were identified by XRD. As shown in Fig. 1, three diffraction peaks were assigned to the hexagonal crystalline-structured ZnO, which match the standard PDF 36-1451. With the decrease of solution concentration, the miscellaneous peaks at 33.051°,

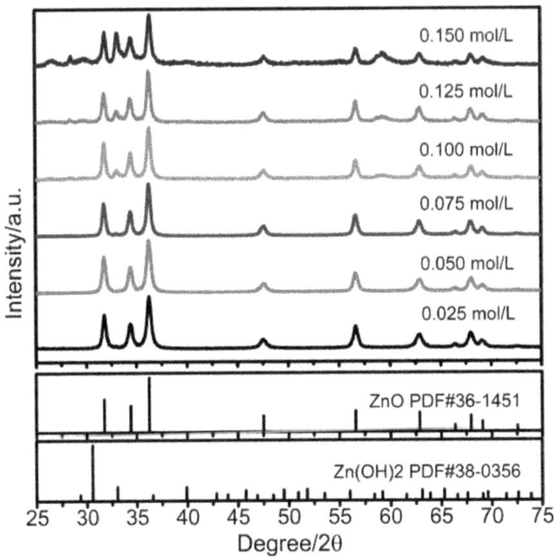

Fig. 1. (Color online) XRD patterns of the ZnO assemblies obtained by different concentrations of the precursor solution.

Table 1. Relevant parameters of ZnO assemblies.

Concentration (mol/L)	(101) FWHM	Crystallite size (nm)	Assembly size (nm)	ρ $(g \cdot cm^{-3})$	γ (mN/m)	Droplet diameter (μm)
0.150	0.414	41.1	870	1.028	71.44	2.875
0.125	0.407	41.8	760	1.023	72.07	2.888
0.100	0.433	39.3	620	1.018	72.16	2.894
0.075	0.430	39.6	650	1.014	72.30	2.899
0.050	0.476	35.7	560	1.009	72.78	2.910
0.025	0.472	36.0	300	1.004	72.90	2.917

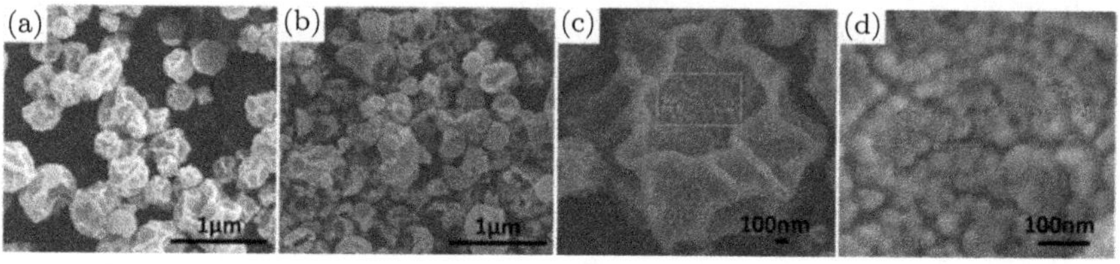

Fig. 2. (Color online) FE-SEM images of the ZnO assemblies obtained by different concentrations of the precursor solution of (a) 0.125 mol/L, (b) 0.05 mol/L (c) and (d) 0.075 mol/L.

39.177° and 58.396° gradually disappeared, which correspond to $Zn(OH)_2$ (PDF 38-0356). The average crystallite sizes of ZnO have been calculated by the Scherrer formula. Table 1 shows that the crystallinity of the particles gradually increases with increasing solution concentration.

Figure 2 shows the FE-SEM images of the ZnO assemblies obtained by different concentrations of the precursor solution. The surface of ZnO assemblies is wrinkled and looks like dried plums, in agreement with Ref. 15. Figures 2(c) and 2(d) show that the particles consist of ZnO. As shown in Fig. 2(d), the size of ZnO nanoparticles is in the range of 33–39 nm, consistent with the results calculated by the Scherrer formula.

FE-SEM images were analyzed to calculate the particle sizes by Image software, as shown in Table 1. With decreasing solution concentration, the microstructure of the powder has no obvious change but the average size of ZnO assemblies decreases from 870 nm to 300 nm, similar to the case in Ref. 14. However, the particles showed better uniform distribution and smaller size compared with those in Ref. 2. These results show that the microstructure and size of the ZnO assemblies could be controlled by the concentration of the precursor solution.

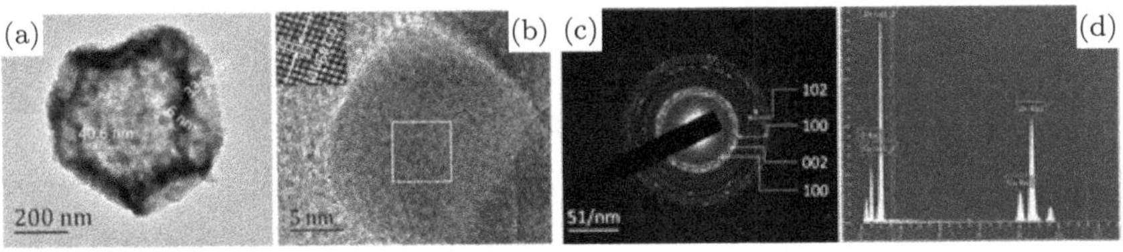

Fig. 3. (Color online) (a) TEM image, (b) HR-TEM image, (c) SAED image and (d) EDS pattern of the ZnO assemblies synthesized by 0.15-mol/L precursor solution. The inset shows the FFTs of the relevant HE-TEM image sections.

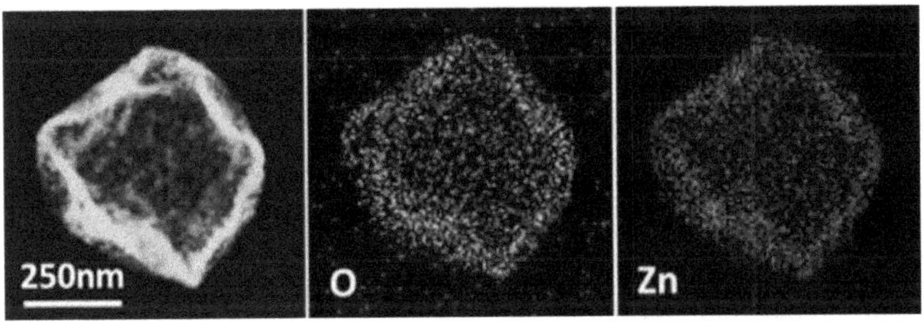

Fig. 4. (Color online) EDS elemental mappings of one ZnO assembly.

Figure 3 shows the TEM analysis of the microstructure of the ZnO assemblies. The ZnO assemblies consisted of nanoparticle ZnO. Figure 3(b) shows the crystal planes of ZnO particles with high crystallinity, consistent with the XRD results. The lattice spacing between the adjacent planes is about 0.25 nm, which matches well with the (101) crystal planes of wurtzite ZnO (0.247 nm) calculated by Bragg's equation. The fast Fourier transform (FFT) patterns are in accordance with the HR-TEM results. Furthermore, the electron diffraction pattern shows the diffraction rings corresponding to the crystallographic orientations of (101), (002), (100) and (102) in a hexagonal wurtzite structure.[14] Figure 3(d) shows the EDS peaks with the composition of Zn and O, which is also proved by the EDS mapping images of ZnO powder in Fig. 4.

Figure 5 shows the schematic route of synthesizing ZnO assemblies. The precursor solution was atomized, evaporated, pyrolyzed and then the ZnO assemblies were fabricated. First, the atomized droplets were transported into zone 1 of the furnace with the carrier gas, and the droplets were evaporated and resolved at 500°C. The water in the crystals was dehydrated

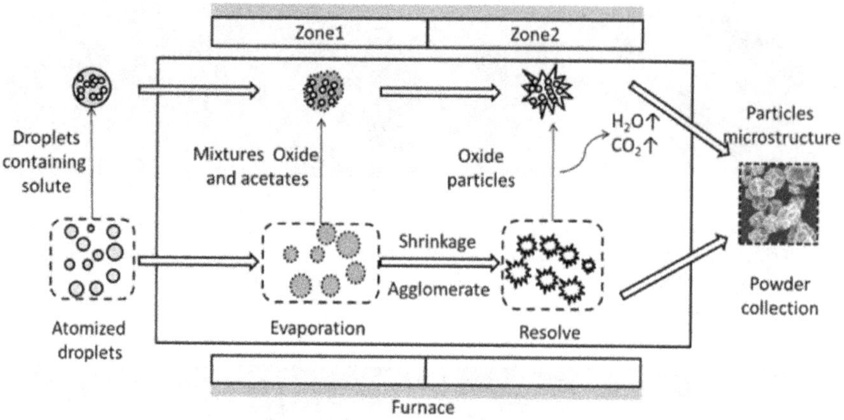

Fig. 5. (Color online) Schematic formation mechanisms for synthesizing ZnO assemblies.

and anhydrous Zn acetate was synthesized at 200°C. Anhydrous Zn acetate was melted at 242°C, and decomposed into ZnO at about 370°C.

Only a small amount of $Zn(CH_3COO)_2$ was pyrolyzed in zone 1 because of the fast air flow. In zone 1, the oxides and acetates were mixed. The mixtures were transported into zone 2 of the furnace, and decomposed into ZnO at 700°C.

During the reaction, the two principle pyrolysis products are water vapor and carbon dioxide. The droplets shrank, aggregated and pyrolyzed because the water vapor and carbon dioxide evaporates, forming the microstructure of dried plum, in accordance with Ref. 16. The droplet diameter (d) is affected by the density (ρ) and the surface tension (γ) of the solution according to the following expression given by Lang[15]:

$$d = 0.34\sqrt[3]{\left(\frac{8\pi \Upsilon}{\rho f^2}\right)}, \tag{1}$$

where f is the applied ultrasonic frequency, which is 1.7 MHz. The droplet diameters were calculated using the equation above and the values were shown in Table 1. With the decreasing concentration of precursor solution, the diameter of the mist droplet increases. The size and concentration of droplet have important effects on the microstructure and size of the obtained ZnO assemblies. In this work, ρ and γ are dependent on the concentration of precursor solution, thus the microstructure and size of the obtained ZnO assemblies can be controlled by the concentration of precursor solution.

4. Conclusions

In this paper, dried plum-like assemblies consisting of ZnO nanoparticles were synthesized by ultrasonic spray pyrolysis with $Zn(CH_3COO)_2 \cdot 2H_2O$. The precursor solution was atomized, evaporated, pyrolyzed and then the ZnO powder was synthesized. With decreasing solution concentration, the microstructure of the powder has no obvious change but the average size of the dried ZnO particles decreases from 870 nm to 300 nm. These results show that the microstructure and size of the ZnO particles can be controlled by adjusting the concentration of the precursor solution. The USP method can be used to synthesize ZnO powder, which shows promise for use in the areas of photocatalysis, LEDs, varistors and gas sensors.

Acknowledgments

This work was supported by the National Natural Science Foundation of China (51402032), Chongqing Youth Science and Technology Talent Cultivation Project (cstc2014kjrc-qnrc40006), Chongqing International Science & Technology Cooperation Program (cstc2015gjhz0003), International Science & Technology Cooperation Program of China (2014DFR50830) and Chongqing University of Arts and Sciences (M2015ME03).

References

1. M. Breedon *et al.*, *Mater. Lett.* **64**, 291 (2010).
2. B. Ebin *et al.*, *Int. J. Miner. Metall. Mater.* **19**, 651 (2012).
3. S. A. Shojaee *et al.*, *J. Mater. Sci.: Mater. Electron.* **21**, 571 (2010).
4. J. Xie *et al.*, *Powder Technol.* **207**, 140 (2011).
5. W. G. Xu *et al.*, *J. Colloid Interface Sci.* **351**, 210 (2010).
6. K. M. Farhan *et al.*, *Sci. Rep. (UK)* **6**, 27689 (2016).
7. S. M. Saleh *et al.*, *J. Environ. Chem. Eng.* **5**, 1219 (2017).
8. A. F. Soares *et al.*, *J. Lumin.* **186**, 135 (2017).
9. C. Byl *et al.*, *J. Sol.-Gel. Sci. Technol.* **83**, 296 (2017).
10. Z. Ye *et al.*, *J. Alloys Compd.* **690**, 189 (2017).
11. N. Shatrova *et al.*, *Mater. Res. Bull.* **86**, 80 (2017).

12. M. Konarova and I. Taniguchi, *Mater. Res. Bull.* **43**, 3305 (2008).
13. O. Milošević, B. Jordović and D. Uskoković, *Mater. Lett.* **19**, 165 (1994).
14. G. Flores-Carrasco *et al.*, *Adv. Mater. Sci. Eng.* **2014**, 780206 (2014).
15. R. J. Lang, *J. Acoust. Soc. Am.* **34**, 6 (1962).
16. T. Li *et al.*, *J. Mater. Chem. A* **3**, 11970 (2015).

https://doi.org/10.1142/9789819818433_0005

Chapter 5

The Influence of Co-doping Strontium and Zirconium on Structure and Ionic Conductivities of $Li_{6+y}La_{3-y}Sr_yNbZrO_{12}$ Solid Electrolytes[#]

Tianhang Zhang[*,†], Yangai Liu[*,||,††], Timothy Daniel Christopher[‡], Saifang Huang[†,**,††], Tao Yang[§], Zhaohui Huang[*], Wei Gao[†], Tilo Söhnel[‡,¶] and Peng Cao[†,¶]

[*]*School of Materials Science and Technology, China University of Geosciences (Beijing), Beijing 100083, P. R. China*

[†]*Department of Chemical & Materials Engineering, University of Auckland, Private Bag 92019, Auckland 1142, New Zealand*

[‡]*School of Chemical Sciences, University of Auckland, Private Bag 92019, Auckland 1142, New Zealand*

[§]*College of Materials & Environmental Engineering, Hangzhou Dianzi University, Hangzhou 310036, P. R. China*

[¶]*MacDiarmid Institute for Advanced Materials and Nanotechnology, Victoria University of Wellington, P. O. Box 600, Wellington 6140, New Zealand*

[||]*liuyang@cugb.edu.cn*

[**]*saifang.huang@auckland.ac.nz*

To enhance the ionic conductivity of lithium garnets, a co-doping strategy was adopted with both strontium and zirconium for $Li_{6+y}La_{3-y}Sr_yNbZrO_{12}$ ($y = 0$, 0.25, 0.5, 0.75 and 1.0) (LLSNZO). By increasing the content of Sr, lithium garnet ceramics maintain cubic structure when y is in the range of 0–0.75. A secondary phase of $SrZrO_3$ appeared in the ceramic when $y = 1.0$. We also studied the cross-section of lithium garnets with silver electrode. Results showed that the density of LLSNZO ceramics increased continuously against the increase of Sr content, while their total ionic conductivity enhanced initially and then reduced, with the maximum reached when $y = 0.5$. It is indicated

[††]Corresponding authors.

[#]To cite this article, please refer to its earlier version published in the *International Journal of Modern Physics B*, Volume 34, 2040006 (2020), DOI: 10.1142/S0217979220400068.

that ionic conductivity of lithium garnets is not only decided by the density but also decided by the concentration and the mobility of Li^+.

Keywords: Solid electrolyte; lithium garnet; ionic conductivity; co-doping strategy; $Li_{6+y}La_{3-y}Sr_yNbZrO_{12}$ (LLSNZO).

1. Introduction

Organic liquid electrolytes in conventional lithium ion batteries (LIBs) are flammable under certain circumstances,[1] which raise a major safety concern in the practical applications of batteries, particularly in the automotive industry. Hence, as a promising replacement of organic liquid electrolytes, solid-state electrolytes attract much attention for the next-generation batteries.

Since in 2003, Thangadurai, Kaack and Weppner firstly reported the structure of lithium garnet $Li_5La_3Nb_2O_{12}$.[2] Researchers have focused on garnet oxides as a potential candidate for inorganic solid-state electrolytes. The lithium garnet oxide electrolytes own excellent electrochemical stability and there is no chemical reaction with lithium anode and cobalt, nickel, manganese oxides at the cathode.[3] Besides, a higher circuit voltage is also probable for lithium garnet oxide electrolytes during charge and discharge progresses due to the wide electrochemical window.[4] However, comparing with liquid electrolytes, inorganic solid-state electrolytes are yet in their fancy, not in practical applications.

Many papers have been published to improve the ionic conductivity of lithium garnets over the past decade. Generally, researchers have paid attention to two main approaches: (1) by increasing the lithium ion concentration or (2) by increasing the mobility of lithium ion. The substitution of La^{3+} by using Ca^{2+}, Sr^{2+} and Ba^{2+} or the substitution of Nb^{5+} by using Ti^{4+}, Zr^{4+} and Hf^{4+} are used to improve the ionic conductivity.[5,6] On one hand, the Li^+ concentration increases so as to maintain the electroneutrality of lithium garnets. On the other hand, the structure has slightly changed after the substitution. Both scenarios favor the mobility of lithium ion in the material, thereby leading to the improvement of its ionic conductivity. The broader lithium diffusion pathway increases the mobility of lithium ion and then improve ionic conductivity of lithium garnet.[7]

In this study, we aimed to synthesize lithium garnet solid electrolytes by co-doping Sr^{2+} for La^{3+} and Zr^{4+} for Nb^{5+} in $Li_5La_3Nb_2O_{12}$. A series of $Li_{6+y}La_{3-y}Sr_yNbZrO_{12}$ lithium garnets (y = 0, 0.25, 0.5, 0.75 and 1.0, terms as LLSNZO thereafter) were synthesized through a conventional solid-state reaction route, and sintered by a pressure-less sintering method. The effect of Sr and Zr co-doping on the structure and properties of the lithium garnets was discussed.

2. Experimental Methods

Li_2CO_3 (Sigma, >99.0% La_2O_3 (Aladdin, >99.9%), $SrCO_3$ (Sigma, >99.0%), Nb_2O_5 (Sigma, >99.0%) and ZrO_2 (Aladdin, >99.99%) were used as starting materials. They were weighed in stoichiometric amounts except that Li_2CO_3 had a 15 wt.% excess amount to compensate the lithium loss at high-temperatures.[8] The mixtures were ball-milled in ethanol using zirconia balls for 12 h, and subsequently calcined in zirconia crucibles to avoid the contamination of aluminum from corundum crucibles,[9] at 900°C for 6 h. Then, powders were pressed under a pressure of 20 MPa into pellets with a diameter of 10 mm. After being pressed, each pellet was covered with mother powders of the same composition to reduce lithium loss, and sintered at 1000°C for 6 h in a muffle furnace.

Phase analysis was performed by powder X-ray diffraction with CuKα radiation (XRD; Rigaku Ultima IV, Japan). Microstructural and morphological studies were performed on an environmental scanning electron microscope (ESEM, FEI Quanta 200F, voltage 20 KeV). Silver paste was coated onto both sides of the sintered pellets and heated at 550°C for 30 min before testing. AC impedance measurements were conducted to measure the ionic conductivity using an impedance analyzer (METEK, PARSTAT MC-2000A, USA) with frequency ranging from 7 MHz to 1 Hz.

3. Results and Discussion

3.1. *Phase analysis*

The phase purity of the $Li_{6+y}La_{3-y}Sr_yNbZrO_{12}$ ceramics with doping concentrations of Sr and Zr ($y = 0$, 0.25, 0.5, 0.75 and 1.0) were analyzed using XRD. As shown in Fig. 1 that after being sintered at 1000°C for 6 h in zirconia crucibles, the cubic structure was maintained in all samples. There is a secondary phase of $SrZrO_3$ in the sample with $y = 1.0$. If the sintering temperature is above 1000°C, the evaporation of lithium is more violent, and consequently another phase $La_2Zr_2O_7$ appeared (data not shown here).

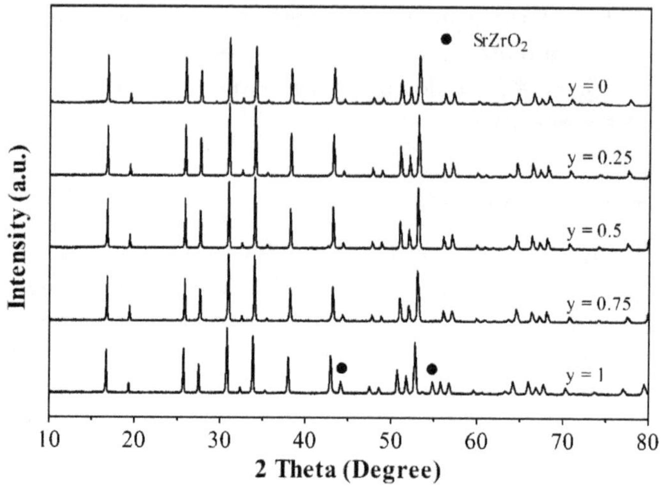

Fig. 1. XRD patterns of the $Li_{6+y}La_{3-y}Sr_yNbZrO_{12}$ ceramics sintered at 1000°C for 6 h.

3.2. *Electrochemical properties*

Typical AC impedance plots measured at room temperature (24°C) for $Li_{6+y}La_{3-y}Sr_yNbZrO_{12}$ ($x = 1.0$; $y = 0$, 0.25, 0.5, 0.75 and 1.0) samples sintered at 1000°C are shown in Fig. 2. It is hard to distinguish between bulk and grain boundary resistances when $y = 0-0.75$. However, there are two obvious semicircles corresponding to bulk and grain boundary resistance in Fig. 2(e) because of the clear grain boundary and the secondary phase $SrZrO_3$. A suitable equivalent circuit was used to fit the impedance plots $(R_b\mathrm{CPE}_b)(R_{gb}\mathrm{CPE}_{gb})\mathrm{CPE}_{\mathrm{int}}$ (where R is the resistance, CPE is the constant phase element, and the subscripts b, gb and int refer to the bulk, grain boundary and interface contribution, respectively).

The variation of total (bulk + grain boundary) ionic conductivity measured at 24°C indicates that, the ionic conductivity enhanced initially with increasing of Sr doped and reached a maximized value of $1.75 \times 10^{-4}\ \mathrm{S \cdot cm^{-1}}$ for $Li_{6.5}La_{2.5}Sr_{0.5}NbZrO_{12}$ from $5.32 \times 10^{-5}\ \mathrm{S \cdot cm^{-1}}$ for $Li_6La_3NbZrO_{12}$. It then reduced with further increasing of the Sr doped, i.e., $2.17 \times 10^{-5}\ \mathrm{S \cdot cm^{-1}}$ for $Li_7La_2SrNbZrO_{12}$.

The LLSNZO garnets with lower doping concentrations showed an increased ionic conductivity when y increases from 0 to 0.5. This is not only because the increased density but also because the increased concentration

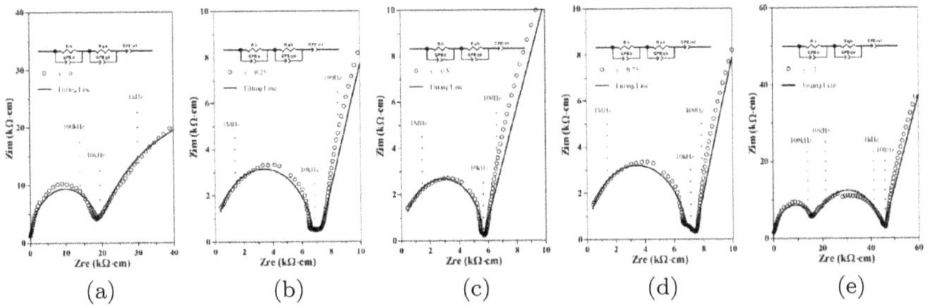

Fig. 2. Typical AC impedance plots measured at room-temperature (24°C) for $Li_{6+y}La_{3-y}Sr_yNbZrO_{12}$ samples sintered at 1000°C using sliver electrodes: (a) $y = 0$, (b) $y = 0.25$, (c) $y = 0.5$, (d) $y = 0.75$ and (e) $y = 1.0$.

and the mobility of Li^+ after doping Sr^{2+} for La^{3+}. With the increasing density of lithium garnets, the negative effects of little pores in those ceramics on lithium ion conductivity weaken.[10] Besides, the substitution of Sr and Zr leads to an enhanced ionic conductivity, which is owing to a higher lithium content in LLSNZO garnets comparing to the undoped $Li_5La_3Nb_2O_{12}$. The substitution of the larger Sr^{2+} (with ionic radius $r = 1.12$ Å, when the coordination number CN = 8) for the smaller La^{3+} ($r = 1.061$ Å, when CN = 8), leads to SrO_8/LaO_8 polyhedron volume expand. The substitution of the larger Zr^{4+} ($r = 0.72$ Å, CN = 6) for the smaller Nb^{5+} ($r = 0.69$ Å, when CN = 6), leads to ZrO_6/NbO_6 polyhedron volume expand. Those changes would lead to an expansion of Li–O polyhedra. This facilitates a slightly broadened Li^+ diffusion pathway and increases the mobility of Li^+, thereby improving the ionic conductivity.[11]

While, the reduction in the ionic conductivity of the lithium garnets with higher substituting concentrations (i.e., $Li_{6.75}La_{2.25}Sr_{0.75}NbZrO_{12}$ and $Li_7La_2SrNbZrO_{12}$) might be attribute to: (1) the excess amount of lithium in LLSNZO composition reduces the disorder degree of lithium and also the mobility of lithium[7] and (2) the secondary phase $SrZrO_3$ between LLSNZO grains in composition increasing grain boundary resistance.

The activation energies (E_a) for the ionic conductivity (σ) were determined by the Arrhenius plots using Eqs. (1) and (2):

$$\sigma T = A\exp\left(\frac{-E_a}{kT}\right), \quad (1)$$

$$\ln(\sigma T) = -\frac{E_a}{kT} + \ln(A), \quad (2)$$

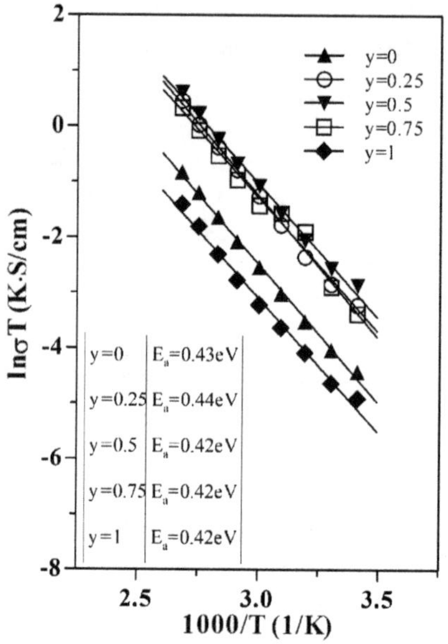

Fig. 3. Arrhenius plots of $Li_{6+y}La_{3-y}Sr_yNbZrO_{12}$ (y = 0, 0.25, 0.5, 0.75 and 1.0) samples measured in the temperature range from 20°C to 100°C.

where A is the pre-exponential parameter; k is the Boltzmann's constant and T is the absolute temperature. The activation energy (E_a) can be calculated from the slope of $\ln(\sigma T)$ versus $1000/T$. The activation energies of all $Li_{6+y}La_{3-y}Sr_yNbZrO_{12}$ ($y = 0$, 0.25, 0.5, 0.75 and 1.0) samples were measured in the temperature range of 20–100°C (shown in Fig. 3). The activation energy of the LLSNZO garnets shows nearly no change, at around 0.43 eV. This indicates that after zirconium doping the activation energy of LLSNZO is quite small, and the strontium doping has seldom influence on the activation energy of $Li_6La_3NbZrO_{12}$.

3.3. *Microstructure*

Figure 4 shows the cross-section of LLSNZO ceramics in different Sr-doping concentrations covered with Ag after being calcined at 550°C for 30 mins. The doping concentration has noticeable effect on the grain size and density. The SEM image of $Li_6La_3NbZrO_{12}$ samples with numerous pores [Fig. 4(a)] clearly reveals the grains sizes are about 5–10 μm. Then the grain size and density of LLSNZO ceramics displayed a continuously increasing trend with the increasing substitution of Sr. It indicates that Sr-doping promotes the

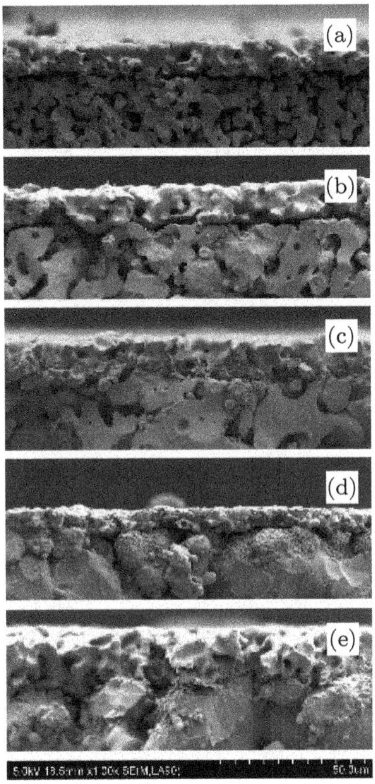

Fig. 4. The cross-section SEM images of $Li_{6+y}La_{3-y}Sr_yNbZrO_{12}$ ceramics with silver electrode calcined at 550°C for 30 min: (a) $y = 0$, (b) $y = 0.25$, (c) $y = 0.5$, (d) $y = 0.75$ and (e) $y = 1.0$.

ceramic sintering process and reduce the sintering temperature of lithium garnets. As a result, it leads to the growth of grain size, the reducing of pores in ceramics and the gradually improved density of ceramics. For instance, the larger grain size (in the range of 20–40 μm) with clear grain boundary was observed in the mostly dense $Li_7La_2SrNbZrO_{12}$ccramics.

4. Conclusions

In this study, we investigated the effect of co-substitution of Sr^{2+} for La^{3+} and Zr^{4+} for Nb^{5+} on the phase structure, sintering and ionic conductivity of lithium garnets ($Li_{6+y}La_{3-y}Sr_yNbZrO_{12}$, LLSNZO). The lithium garnets maintained the cubic structure by substituting strontium with y value up to 0.75. While $y = 1.0$, there was a secondary phase of $SrZrO_3$ in the material. Among the Sr and Zr co-doped lithium garnets, $Li_7La_2SrNbZrO_{12}$ (i.e., $y = 1.0$) ceramics have the highest relative density 86.67% but the largest grain size 20–40 μm and the most obvious grain boundary. $Li_{6.5}La_{2.5}Sr_{0.5}NbZrO_{12}$ (i.e., $y = 0.5$) ceramics have the optimistic ionic conductivity 1.75×10^{-4} S · cm^{-1} enhanced from 5.32×10^{-5} S · cm^{-1} of $Li_6La_3NbZrO_{12}$. Nevertheless, the activation energy of lithium garnets keeps similar with increasing of Sr content, with approximately 0.43 eV for all these garnet compounds.

Acknowledgments

We acknowledge the support from the Science for Technological Innovation (SfTI) National Science Challenge (NSC), New Zealand. T. Zhang gratefully acknowledges the scholarship support from the China Scholarships Council (CSC) for his study at the University of Auckland.

References

1. S. Ohta *et al.*, *J. Power Sour* **265**, 40 (2014).
2. V. Thangadurai, H. Kaack and W. J. F. Weppner, *J. Am. Ceram. Soc.* **86**, 437 (2003).
3. S. Ohta *et al.*, *J. Power Sources* **202**, 332 (2012).
4. S. Ohta *et al.*, *J. Power Sources* **238**, 53 (2013).
5. S. Ohta, Y. Kihira and T. Asaoka, *Front. Energy Res.* **4**, 1 (2016).
6. S. G. Kang and D. S. Sholl, *J. Phys. Chem. C* **118**, 17402 (2014).
7. W. G. Zeier, *Dalton Trans.* **43**, 16133 (2014).
8. S. Narayanan *et al.*, *J. Electrochem. Soc.* **162**, A1772 (2015).
9. T. Thompson *et al.*, *J. Mater. Chem. A* **2**, 13431 (2014).
10. L. Li *et al.*, *J. Sol.-Gel. Sci. Technol.* **83**, 660 (2017).
11. D. W. Wang *et al.*, *Chem. Mater.* **27**, 6650 (2015).

https://doi.org/10.1142/9789819818433_0006

Chapter 6

A High-Performance Graphene Based Asymmetric Supercapacitor[#]

Syeda Wishal Bokhari[*,†], Ahmad Hassan Siddique[‡,§], Harshpreet Singh[*], Muhammad Dilawer Hayat[*], Shenmin Zhu[†] and Wei Gao[*,¶]

[*]*Department of Chemical and Materials Engineering, University of Auckland, Private Bag 92019, Auckland 1142, New Zealand*

[†]*State Key Laboratory of Metal Matrix Composites, School of Material Science and Engineering, Shanghai, Jiao Tong University, Shanghai 200240, P. R. China*

[‡]*Key Laboratory of Graphene Technologies and Applications of Zhejiang Province, Ningbo Institute of Materials Technology and Engineering (NIMTE), Chinese Academy of Sciences, Ningbo 315201, P. R. China*

[§]*University of Chinese Academy of Sciences, 19 A Yuquan Rd., Shijingshan District, Beijing 100049, P. R. China*

[¶]*w.gao@auckland.ac.nz*

We present the design of a high-performance 2 V asymmetric supercapacitor made of graphene-based hybrid nanocomposites in 1 M KOH electrolyte. rGO-CNT-FeOOH (GC-F) was used as the cathode and rGO-CNC-MnO_2 (GC-M) as the anode. Both composites well balanced each other in the asymmetric device and gave a high electrochemical performance. The energy density of this ASC is 66.6 W h kg^{-1} at 3.8 kW kg^{-1} power density with an excellent capacitance retention of 88% after 3000 galvanostatic charge-discharge cycles.

Keywords: Carbon based materials; Nanocomposites; Supercapacitors.

[¶]Corresponding author.

[#]To cite this article, please refer to its earlier version published in the *International Journal of Modern Physics B*, Volume 34, 2040007 (2020), DOI: 10.1142/S021797922040007X.

1. Introduction

Supercapacitors are an attractive energy storage system due to their high power density, long life cycle and rapid charge-discharge rates. They have a simple working mechanism in which an excess or deficiency of ions take place during charging and the reverse takes place during discharging.[1]

Supercapacitors are divided into three major categories based on their energy storage mechanisms: (1) electrochemical double layer capacitors (EDLCs) which have an electrostatic mode of storing charges by making a Helmholtz double layer on the surface of electrode materials; (2) pseudo capacitors (PCs) which have a Faradic mode of storing charges by the electrochemical reactions through the electrode materials and (3) hybrid/asymmetric supercapacitors (HSCs/ASCs) which combine two distinctive energy storage modes together in a single device and output a better performance than the parent systems.[2] For asymmetric supercapacitors, one electrode comes from the double layer capacitor and the other comes from the pseudocapacitors. Whereas, hybrid supercapacitors combine one electrode from the supercapacitors and the other is a battery-type electrode. The third type of supercapacitors has been so far the best energy-power profile.[3,4]

Carbonaceous electrodes are preferred for the EDCLs whereas metal oxides, (TMOs) are the favored electrodes for the PCs.[5,6] Carbonaceous materials offer a high surface area for charge storage, however their low specific capacitance is a major hurdle in their industrial applications. On the other hand, TMOs do offer a very high specific capacitance (1000–1800 F g^{-1}), but they have a low conductivity and low cyclic stability which greatly affects their rate performance.[7]

Therefore, one of the most effective approach is to design hybrid electrodes by using carbon-based conductive matrix and TMOs-based electroactive component. The resultant nanocomposites show an improved conductivity and cyclic stability, high specific capacitance and rate performance.[8]

The high specific surface area (SSA), great vulnerability of graphene to the ion intercalation, superior capacitance retention, long cyclic life and the tunable structural configuration make graphene an ideally suited candidate for supercapacitor application.[9] Its electrochemical performance can further be enhanced by making it hybrid with TMOs to add to its superior properties.[10]

Herein, we design an asymmetric system using two hybrid nanocomposites in 1 M KOH electrolyte. The device gave a high specific capacitance of

120.1 F g^{-1} at 1 A g^{-1}, 88% capacitance retention after 3000 cycles and a high energy density of 66.6 W h kg^{-1} at 3.8 kW kg^{-1} power density. This scheme can be used to design asymmetric system based on graphene hybrid electrodes with a superior performance in aqueous electrolytes.

2. Experimental Methods

Two graphene-based nanocomposites were prepared by simple hydrothermal method given in our previous report.[7] Briefly, two graphene-based conductive matrix rGO-CNT and rGO-CNC are prepared by the hybridization of graphene with CNT and CNC to avoid the agglomeration of graphene sheets. Then, FeOOH and MnO_2 are deposited on to the matrix and hydrothermally treated at 120°C for 4 h. All assemblies are conducted *in situ* via hydrothermal method to ensure maximum contact between the nanocomposite components. Finally, the prepared nanocomposites are carbonized in an inert atmosphere to activate the matrix and increase the conductivity.

The electrochemical performance has been studied using cyclic voltammetry (CV), galvanostatic charge-discharge curves (GCD) and electrochemical impedance spectroscopy (EIS) on a CHI-660E electrochemical analyzer (Chenhua Instrument Inc., Shanghai). The electrodes were prepared by grinding the nanocomposites, carbon black and PVDF binder in 80:10:10 and a thick slurry was made using NMP as a solvent. Ni-foam was used as a current collector and the slurry-coated electrodes were dried in an oven at 90°C for 1 h and 60°C overnight to break the binder and dry the electrodes. The active mass ratio of both working electrodes was optimized in order to achieve the maximum electrochemical performance. On average, the active mass on the electrodes was kept at 2.3 ± 0.1 mg. The specific capacitance of the ASC has been calculated by GCD curves using Eq. (1). The energy and power densities are calculated using Eqs. (2) and (3), respectively.

$$C = \frac{I \times t}{m \times V}, \tag{1}$$

$$E = \frac{1}{2} C V^2, \tag{2}$$

$$P = \frac{E}{\Delta t}, \tag{3}$$

where C is the specific capacitance of the material (F g^{-1}), I is the charge-discharge current density (A), t is the charge-discharge time (second), m is the mass of active material on each electrode (g), V is the voltage window, E is the energy density (W h kg^{-1}), P is the power density (kW h kg^{-1}) and Δt is the discharge time in seconds.

3. Results and Discussions

The electrochemical performance of the designed asymmetric system has been given in Fig. 1. The CV curves in Fig. 1(a) shows a typical rectangular shape which are characteristic to the conductive nature. Some small peaks can be seen in oxidation and reduction peaks coming from the pseudocapacitive character added by the TMOs. The shape of CV curves

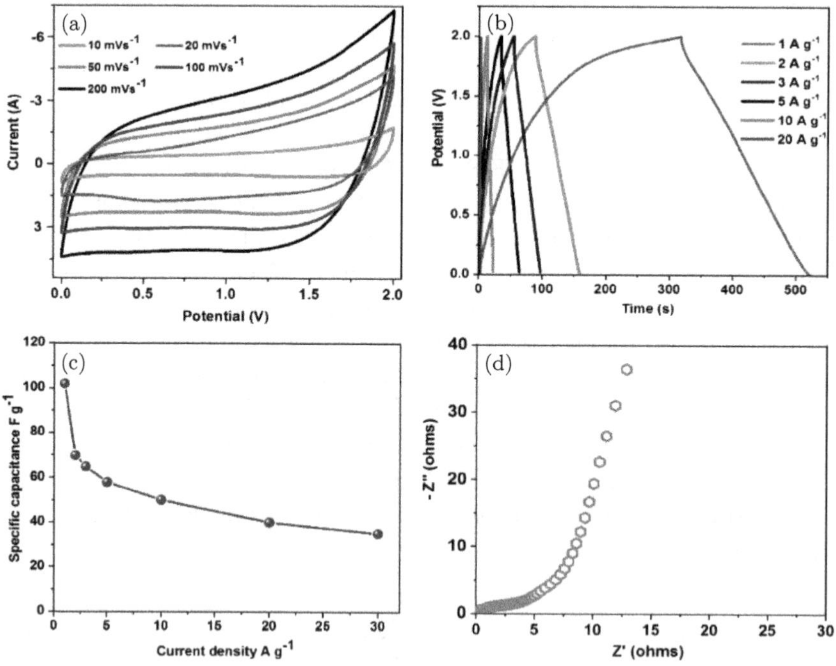

Fig. 1. (Color online) Electrochemical performance of GC-F//GC-M asymmetric supercapacitor: (a) CV, (b) GCD curves of the as assembled asymmetric supercapacitor using GC-M as a cathode and GC-M as an anode, (c) specific capacitance at various current densities and (d) EIS spectra of the asymmetric supercapacitor after 3000 charge-discharge cycles.

remains rectangular at both high and low scan rates, indicating the good rate performance of ASC.

The GCD curves given in Fig. 1(b) represents a sharp IR drop, demonstrating the high charge-discharge property and conductive nature in accordance with the CV curves. The specific capacitance calculated using GCD is 120.1 F g^{-1} at the current density of 1 A g^{-1} even at higher current density of 20 and 30 A g^{-1}, and the ASC maintained a capacitance as high as 49 and 31 F g^{-1}. This can further be seen in the rate performance of the ASC given in Fig. 1(c).

Furthermore, the EIS has been carried out to study the resistance. As displayed in Fig. 1(d), there is a very small semicircle in high frequency region, representing that the ASC has a very small charge transfer resistance ($\sim$1.2 Ω). In the high frequency region, a straight line is observed which refers to the conductive behavior and a low contact resistance. This result is in agreement with the CV and GCD observation, and further justifies the high conductance and in turn the good electrochemical nature of ASC.

The cycling stability of the ASC was recorded for 3000 galvanostatic charge-discharge cycles. The asymmetric supercapacitor has shown a good capacitance retention of 88% after 3000 GCD cycles as shown in Fig. 2.

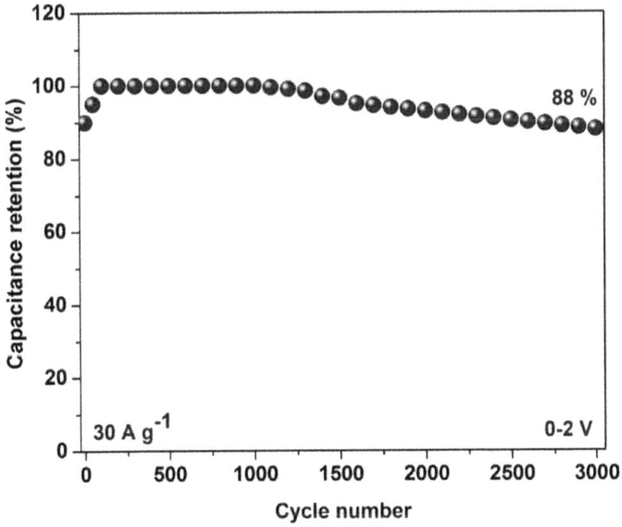

Fig. 2. (Color online) Cycling performance of the GC-F//GC-M asymmetric supercapacitor over 3000 charge-discharge cycles recoded at an ultra-high current density of 30 A g^{-1}.

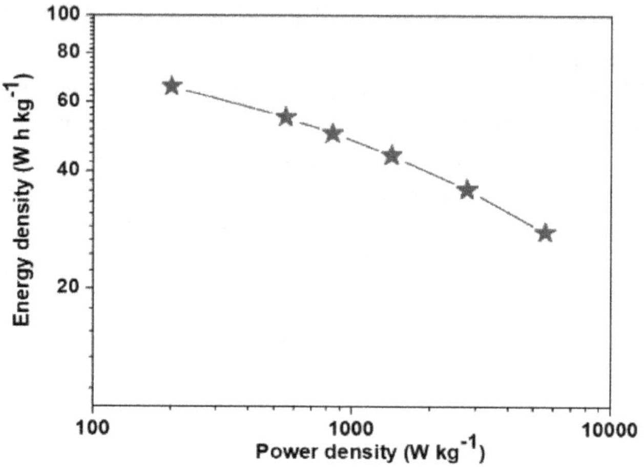

Fig. 3. (Color online) Ragone plot. Energy and power density of the designed ASC.

From 1st to 100th cycles, the capacitance was around ~91%, which then increased to 100%. This can be explained as the electrolyte takes time to properly wet the pore wall and contact with the electrode interfaces. After the intimate contact between the electrode and electrolyte interfaces, the electrodes show their full electrochemical potential. This test further suggests that the electrodes should be immersed in electrolyte for some time to ensure well contact between electrode and electrolyte interfaces. The "*wetting*" pore wall process avails the maximum electrochemical potential of the electrode materials in asymmetric supercapacitor configuration.

The energy and power density of the ASC has been calculated using Eqs. (2) and (3), respectively. The highest energy density of the ASC is 66 W h kg^{-1} at an ultra-high power density of 3.8 kW kg^{-1} (Fig. 3). Hence, it can be concluded that the asymmetric devices incorporating hybrid and well-structured electrode are capable of delivering excellent electrochemical performances and cycling stability originating from the synergistic relationship between the two well-ordered and highly stable hybrid nanocomposites.

4. Conclusion

To enhance the electrochemical performance of the graphene-based matrix, we coupled them with transition metal oxides (TMO) which functioned as an electroactive component. The composite was then used as electrode-based materials for the constructing a 2 V asymmetric supercapacitor with a high electrochemical performance originating from the synergistic effect. The ASC has delivered a high energy density of 66.6 W h kg^{-1} at 3.8 kW kg^{-1} power with 88% capacitance retention after 3000 charge-discharge cycles. The high specific capacitance, excellent capacitance retention and high energy and power densities of this system suggest its great potential for asymmetric supercapacitors. This scheme can be useful in designing high capacity ternary electrode materials for supercapacitor applications.

Acknowledgments

S.W. Bokhari would like to thank Shanghai Jiao Tong University for providing laboratory resources to conduct this experiment and to the University of Auckland for providing Doctoral Scholarship and resources for writing up this manuscript.

References

1. A. Borenstein *et al.*, *J. Mater. Chem. A* **5**, 12653 (2017).
2. S. W. Bokhari *et al.*, *RSC Adv.* **7**, 18926 (2017).
3. D. P. Dubal *et al.*, *Chem. Soc. Rev.* **44**, 1777 (2015).
4. W. Zuo *et al.*, *Adv. Sci.* **4**, 1600539 (2017).
5. A. C. Forse *et al.*, *J. Am. Chem. Soc.* **138**, 5731 (2016).
6. B. Gao *et al.*, *Thin Solid Films* **584**, 61 (2015).
7. S. W. Bokhari *et al.*, *Mater. Lett.* **220**, 140 (2018).
8. P. Geng *et al.*, *Adv. Energy Mater.* **8**, 1703259 (2018).
9. O. Ke and J. Wang, *J. Materiomics* **2**, 37 (2016).
10. X. Li and L. Zhi, *Chem. Soc. Rev.* **47**, 3189 (2018).

https://doi.org/10.1142/9789819818433_0007

Chapter 7

Nanoshell Si/Cu Composites as Anode Materials for Lithium Ion Battery[#]

Manshu Zhang*, Lian Shen† and Yangai Liu‡

School of Materials Science and Engineering,
China University of Geosciences (Beijing),
29 Xueyuan Road, Beijing 100083, P. R. China
**zhmans@126.com*
†sgl1028@sina.com
‡liuyang@cugb.edu.cn

The chemical plating methods were used to prepare nanoshell Si/Cu composites. The methods delivered a high initial discharge and charge capacity of 1312.1 and 1008.8 mAh/g, respectively. The electrochemical properties of the composites retained about 662.6 mAh/g after 100 cycles. The Cu coating improved the electrical conductivity but hindered the direct contact of the Si particles and the electrolyte, which suppressed the SEI growth.

Keywords: Shell; Si/Cu composites; anode material.

1. Introduction

Lithium ion batteries (LIBs) are considered to have the most potential energy storage configuration, due to their high energy and power capacity, long cycle life, low cost and high security.[1] Since the development of high-performance electric vehicles and energy-storage power stations, current commercial LIBs with graphite anodes (theoretical capacity 372 mAh g^{-1}) are unable to satisfy these application requirements.[2] Silicon (Si) is a promising anode material for LIBs, due to its high theoretical capacity of 4200 mAh g^{-1} via forming $Li_{4.4}Si$. However, Si suffers from a large volume expansion during lithiation and delithiation process, resulting in

‡Corresponding author.

#To cite this article, please refer to its earlier version published in the *International Journal of Modern Physics B*, Volume 34, 2040008 (2020), DOI: 10.1142/S0217979220400081.

severe particle pulverization and fast capacity fading, which hinders its commercialization process.[3]

Si-M composites (M = nonactive materials, e.g., carbon, TiN, TiC, SiC, Fe, Ni, Cu, etc.) have been developed to reduce the severe particle pulverization and fast capacity fading. Though its capacity decreases when M is introduced into the Si matrix, its cycle performance improves to a certain extent.[4] Si/Copper (Cu) composites attract great attention due to their excellent electrical conductivity, high diffusivity and good solubility of Cu into Si, which lowers the polarization of the electrode. The good ductility of Cu guarantees the buffer space for the large volume expansion. Cu alloys with Si to form Cu_3Si and $Cu_{15}Si_4$ are both electronic conductors. The high stability of Cu as an anode avoids the side reactions between the Si and the electrolyte.[5–7]

Various synthetic routes have been established to prepare the Si/Cu composites. Metallic Cu has successfully been deposited onto the surface of micrometer silicon particles by Zhong and Shen *et al.*, which efficiently improved the cycle performance of the silicon anode.[9] He and Arreaga-Salas *et al.* investigated the electrochemical properties of the Si-Cu films, where the appearance of the Si/Cu alloy buffers the large volume expansion of the silicon anode.[10] Guo and Worsley *et al.* proposed a novel concept that uses Cu-Si nanocable arrays that are directly anchored on a current collector.[11] This promotes both the cycling stability and the high rate capability.[8] Murugesan and Yang *et al.* reported a polyol reduction method to synthesize Si/Cu composite, where Cu-coated amorphous Si:H particles were successfully obtained with excellent electrochemical performance.[11,12]

In this paper, Cu was deposited onto the surface of nanosilicon by using a typical chemical plating method. An improved plating solution was proposed. The nano Si/Cu composites were characterized by X-ray diffraction (XRD), scanning electron microscopy (SEM) and electrochemical impedance spectroscopy (EIS). These results demonstrated that the existence of the Cu coating enhanced the charge transfer kinetics. The cycle performance and the rate capability of the nano Si/Cu composites was improved.

2. Experimental Methods

Nanosilicon was pretreated with a typical procedure in order to achieve superior interfacial properties. Silicon was ultrasonically dispersed in acetone for 30 min to remove the surface impurities. After the full volatilization of acetone, further sensitization and activation were performed. The as-prepared Si particles slowly dispersed into copper plating solution with 15 g/L of $CuSO_4 \cdot 5H_2O$ in 100 mL of methyl alcohol. NaOH was used as a buffer agent to control the pH, methanol was used as a reductant and eidetic acid was used as a complexing agent. The reactions were performed at room temperature. The copper ions were reduced due to the autocatalytic process of the pretreated Si particles, and then deposited onto its surface. A stronger binding force between the Si and the Cu ions was achieved by sintering at high temperatures (400–600°C) to form Si/Cu alloy. The nano Si/Cu composites were obtained after cooling.

3. Results and Discussion

The XRD patterns of the as-prepared samples are presented in Fig. 1. The characteristic diffraction peaks were observed at 28°, 47°, 56°, 69° and 76°, corresponding to the (111), (220), (311), (331) and (422) crystalline planes of Si (JCPDS No. 27-1402), respectively. The as-prepared sample had peaks at 43°, 50° and 74°, which were indexed to copper (JCPDS No. 27-1402) and indicated the existence of metallic copper. This denoted that copper had been successfully coated on nanosilicon particles, although the copper

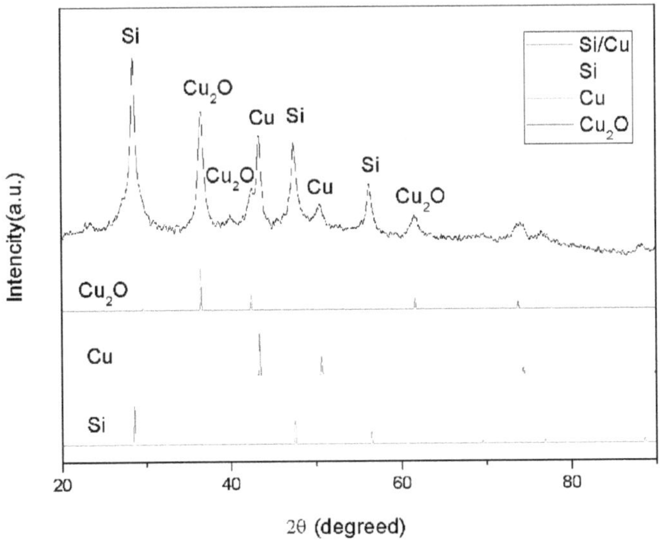

Fig. 1. XRD spectra of Si/Cu composites.

Fig. 2. SEM images of the pure silicon particles (a) SEM images of nano-Si, (b) SEM images of the Si/Cu composites and (c,d) TEM images of the Si/Cu composites.

was oxidized to Cu_2O during the copper plating. Theoretically, the Cu_2O would not affect the cycling process because the Cu_2O will be reduced to Cu during the initial discharge process, only the Coulombic efficiency can be influenced during the first cycle.

Figures 2(a) and 2(b) show the SEM images of the pure silicon and silicon Si/Cu composites, respectively. Most of the nanosilicon particles reunited to form a particle shape. The particle size was more than a dozen to hundreds of nanometers, the particle surface was not smooth, there were many edges, and had an irregular spherical shape. After the copper coating, the particle size remained nearly the same, but the silicon particle surface did not seem to have the edge, particles on the edge of the line were smooth. TEM was used to observe the microscopic structure of the Si/Cu composites [Figs. 2(c) and 2(d)]. Particles are observed in the picture surface with a thin layer between 2 nm and 10 nm, due to agglomeration of the particle's morphology. There were a couple of particle surface cladding layers.

During the initial scan process in Fig. 3(a), the diffraction peak located in 0.01 V in the reduction curve corresponds to the lithium ion embedding process, which means the formation of the Li-Si alloy. The oxidation peaks located in 0.35 V and 0.57 V is corresponding to the lithiation process, and

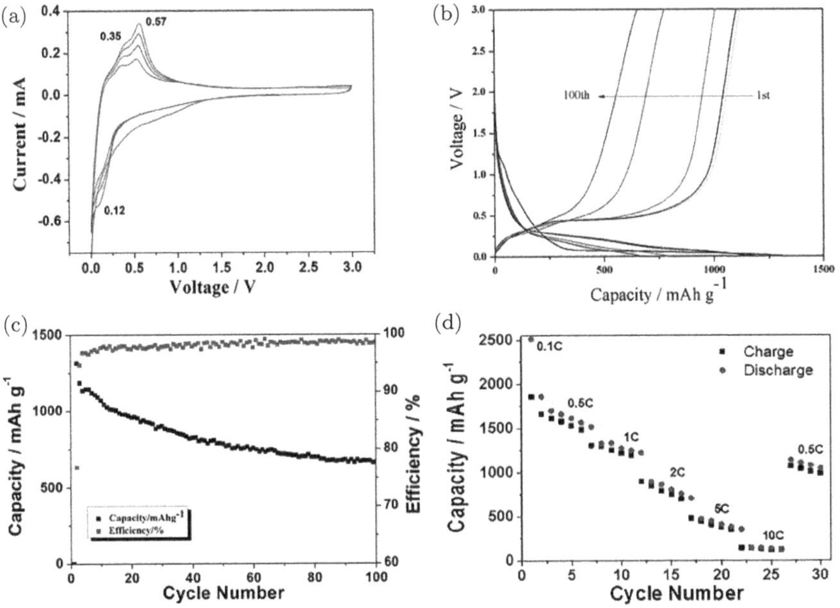

Fig. 3. (a) The cyclic voltammetry (CV) curves of the Si/Cu composites, (b) the charge and discharge curves, (c) the long-term circulation performance and (d) the ratio performance.

the reduction peaks in 0.12 V is corresponding to the lithiation intercalation process. The curve has a high repeatability after the first cycle, which means the cycling stability of the materials is pretty good. Figure 3(b) shows that the discharge and charge specific capacity of the sample in the first cycle were 1312.1 mAh/g and 1008.8 mAh/g at 0.1 A/g, respectively. The Coulombic efficiency in the first cycle was 76.88%. And after 100 cycles the reversible capacity is 662.6 mAh/g [Figure 3(c)]. Figure 3(d) is the rate performance of the materials, which shows that the rate performance of the materials is not pretty good, the reversible specific capacity can maintain above 1000 mAh/g when the rate is 0.1, 0.5, 1 C, and the capacity fading is a little quick, which can be ascribed to the high ratio of the copper, resulting in the diffusion process of the Li^+ was limited.

Figure 4 shows that the impedance spectra had characteristics of the two semicircles with a slash added. The high-frequency semicircle corresponded to the lithium ion diffusion impedance in the SEI film. The intermediate frequency semicircle corresponded to the charge transfer

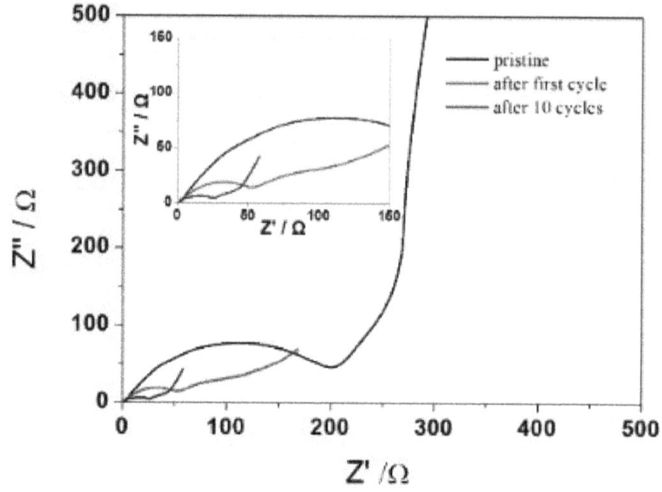

Fig. 4. EIS of the Si/Cu composites.

impedance electrode. The diagonal lines were associated with lithium ion diffusion in the electrode active material. The original Si/Cu composites corresponded to the charge transfer impedance. The first half of the second round corresponded to the Si particles, the Cu and the interface between the electrolyte impedance, and the slash with Warburg diffusion impedance. As the circulation of materials of reaction kinetics continued, the diffusion impedance, the charge transfer resistance and the spread of Warburg impedance decreased accordingly. This showed that the Cu on Si nanoparticles surface cladding layer helped to improve the electronic conductivity of the silicon-based anode and the reaction kinetics. The Cu layer effectively prevented direct contact between the Si particles and the electrolyte, reduced the incidence of adverse events and inhibited the growth of the SEI film. The cycle performance and the rate performance of silicon-based anode were improved.

4. Conclusions

The chemical copper plating method was successfully used in the silica nanoparticles surface cladding layer, where a uniform layer of copper on Si/Cu nanocomposite was achieved. In the first cycling process, the discharge and charge specific capacity were 1312.1 mAh/g and 1008.8 mAh/g at the current density of 0.1 A/g, and the first cycle Coulombic efficiency is 76.88%; after 100 cycles the reversible capacity can still retain at 662.6 mAh/g. The existence of the Cu layer effectively improved the silicon-based electronic conductivity of the composite materials, and successfully avoided the contact between Si particles and the electrolyte, thus improving the cycle performance and the rate of silicon-based anode performance.

Acknowledgments

This work was supported by "the Fundamental Research Funds for the Central Universities" (No. 51772278).

References

1. J. M. Tarascon, *CR Chim.* **8**, 9 (2005).
2. G. Kucinskis, *J. Power Sour.* **240**, 66 (2013).
3. J. R. Szczech, *Energy Environ. Sci.* **4**, 56 (2011).
4. R. N. Seefurth, *J. Electrochem. Soc.* **124**, 1207 (1977).
5. T. Yang, *J. Alloy Compd.* **735**, 1079 (2018).
6. N. A. Liu, *Sci. Rep.* **3**, 181 (2013).
7. S. C. Lai, *J. Electrochem. Soc.* **123**, 1196 (1976).
8. C. K. Chan, *ACS Nano* **4**, 1443 (2010).
9. DE Arreaga-Salas, *J. Phys. Chem. C* **116**, 9072 (2012).
10. L. Shen, *Chem. Eur. J.* **21**, 8491 (2015).
11. M. A. Worsley, *Chem. Commun.* **20**, 4840 (2010).
12. T. Yang, *Energy Storage Mater.* **17**, 374 (2019).

https://doi.org/10.1142/9789819818433_0008

Chapter 8

Surface-Enhanced Raman Scattering (SERS) by Ag Nanoparticles on Anodized TiO_{2-X} Nanotubes[#]

Ting Zhang[*], Junzhi Ye[*,†], Rakesh Arul[*,†], Tingxuan Yang[*], Yixuan Wang[*], Xiaoyu Yue[*], Marcel Schaefer[‡], Cather Simpson[†], Michél K. Nieuwoudt[†], Saifang Huang[*], Shanghai Wei[*] and Wei Gao[*,§]

[*]*Department of Chemical & Materials Engineering, The University of Auckland, Private Bag 92019, Auckland 1142, New Zealand*

[†]*The Photon Factory, School of Chemical Sciences, The University of Auckland, Auckland 1142, New Zealand*

[‡]*Department of Mechanical Engineering, Auckland University of Technology, Auckland, New Zealand*

[§]*w.gao@auckland.ac.nz*

In this work, Ag nanoparticles were deposited in the TiO_{2-X} nanotube films through magnetron sputtering to form Ag-TiO_{2-X} nanocomposite. Scanning electron microscopy (SEM) was used to study microstructure and surface morphology. Rhodamine B (RhB) was deposited on the surface of Ag-TiO_{2-X} substrate for Raman spectroscopy test. The size of Ag nanoparticles can be controlled by changing the magnetron sputtering time. The result indicated that sputtering for 30 s obtained the highest Raman enhancement.

Keywords: Titanium oxide substrate; anodization; Ag nanoparticles; SERS.

1. Introduction

Recently, research on the noble metal coated semiconductor-based surface-enhanced Raman scattering (SERS) has attracted much attention, which has wide applications on medical diagnostics, security, plant and animal disease, and within the chemical industry.

[§]Corresponding author.

[#]To cite this article, please refer to its earlier version published in the *International Journal of Modern Physics B*, Volume 34, 2040009 (2020), DOI: 10.1142/S0217979220400093.

TiO_{2-X} is one of the most widely investigated semiconductor material in recent years. Its applications include photocatalysts,[1] various batteries[2] and solar cells.[3–5] The oxygen vacancy and Ti^{3+} in the crystal structure can reduce the recombination of electrons and holes and increase the conductivity of TiO_{2-X}.[6] These unique properties of TiO_{2-X} led to a widespread use as a SERS substrate.

It has been known that the SERS enhancement of molecules absorbed on the TiO_{2-X} is ascribed to the electromagnetic mechanism (EM) caused by Ag nanoparticles and charge-transfer (CT) effect coming from TiO_{2-X} itself. EM is attributed to the dominant contribution of SERS Enhancement because of the localized surface plasma resonance (LSPR) effect of Ag nanoparticles on the surface of TiO_{2-X} nanotubes.

However, the relationship between the size of Ag nanoparticles coated on the anodized TiO_{2-X} nanotubes and Raman enhancement factor has not been clear, which is important for further enhancement of Raman signals. In this study, the different amounts of Ag nanoparticles are coated on the surface of TiO_{2-X} surface by magnetron sputtering. TiO_{2-X} nanotubes are made by two steps of anodization as the SERS-active substrates. We investigated the photo-induced SERS mechanism influenced by the size of Ag nanoparticles.

2. Experiment

2.1. *Sample preparation*

The highly ordered TiO_{2-X} nanotubes were prepared by a two-step process of electrochemical anodization on the thin titanium foil (>99.6%Ti, $40 \times 25 \times 0.5$ mm), which was mechanically polished and ultrasonically cleaned in distilled water. The first anodization process was carried out under 60 V for 25 h in 250 mL electrolyte of ethylene glycol containing 0.25 wt.% NH_4F and 2 vol.% distilled water. The obtained TiO_2 nanotube layer was removed and the second anodization was performed on the anodized Ti foil in the same electrolyte at 80 V for 5 min.

The TiO_2 thin film was then washed in ethanol and dried in N_2. Finally, TiO_{2-x} nanotubes were obtained by annealing at 450°C in Ar for 1 h.

Ag-deposited TiO_{2-x}(Ag-TiO_{2-x}) nanocomposites were prepared by magnetron sputtering using pure Ag target under an argon atmosphere. TiO_{2-X} substrate was coated by Ag nanoparticles with different size controlled by the different sputtering time (20, 25 and 30). After sputtering, all samples conducted dewetting treatment in Ar for 1 h at 250°C.

2.2. *Sample characterization*

The surface morphology of the sample was studied by scanning electron microscope (HITACHI SU-70) under an acceleration voltage of 5 KV. Raman spectra were obtained using a RM 1000 Raman instrument equipped with an air-cooled Argon ion 488 nm excitation laser. The acquisition time for each accumulation was 60 s. For the SERS measurement, 5 μL 1.0×10^{-5} M Rhodamine B was dropped on each of the Ag-TiO_{2-X} nanocomposite substrates.

3. Results and Discussion

Morphologies of anodized Ti film with different Ag size are given in Fig. 1. The sputtering times are 20, 25 and 30 s, respectively. It can be seen that the size of Ag nanoparticles becomes larger with increasing sputtering time to 30 s, which can be attributed to the aggregation and growth of Ag nanoparticles. Figure 2 shows the Ag nanoparticles distribution on the TiO_{2-X} substrate with increasing sputtering time. The percentage of particles ranging from 60 nm to 80 nm is 2.1% in 20 s, 18.6% in 25 s and 18.8 in 30 s, as shown in Fig. 2.

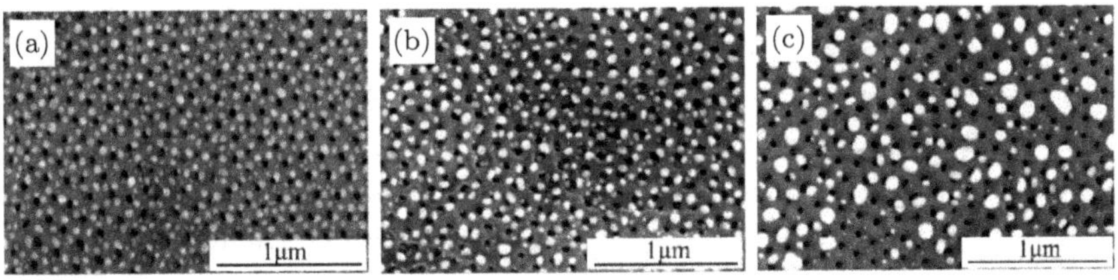

Fig. 1. SEM images of Ag nanoparticles deposited by magnetron sputtering for different times: (a) 20 s, (b) 25 s and (c) 30 s.

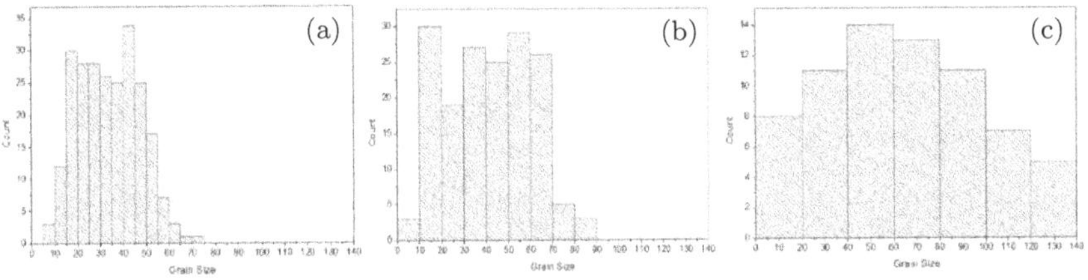

Fig. 2. (Color online) Particle size distribution of different magnetron sputtering time: (a) 20 s, (b) 25 s and (c) 30 s.

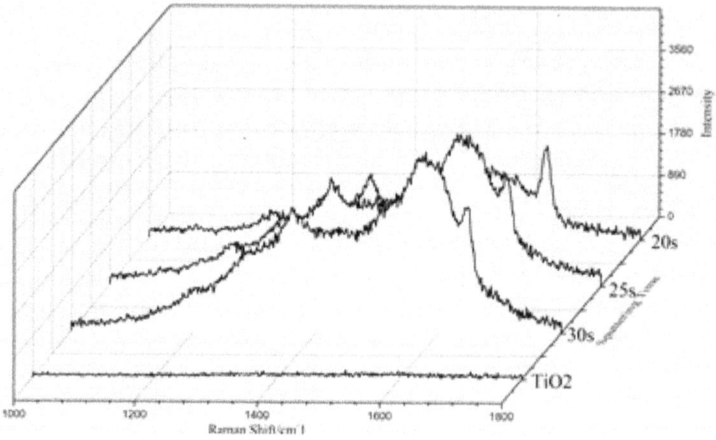

Fig. 3. Raman spectra of RhB from the Ag-TiO_{2-X} nanocomposite substrate with different Ag magnetron sputtering time.

To investigate the dependence of SERS enhancement on the size and density of Ag nanoparticles, SERS spectra of RhB absorbed on different Ag-TiO_{2-X} samples have been tested and the results are shown in Fig. 3. Small size and low density of Ag nanoparticles are not the optimum conditions to be excited by a laser. The Ag-TiO_{2-X} sample prepared by sputtering for 30 s exhibits the highest SERS intensity, which can be attributed to the most Ag nanoparticles ranging from 60 nm to 80 nm as shown by SEM images and Ag nanoparticle contribution. To estimate the enhancement ability of Ag-TiO_{2-X} nanocomposite, the enhancement factor can be calculated by using the following equation:

$$EF = \left(\frac{I_{\text{SERS}}/N_{\text{SERS}}}{I_{\text{bulk}}/N_{\text{bulk}}}\right) = \left(\frac{I_{\text{SERS}}/I_{\text{bulk}}}{C_{\text{SERS}}/C_{\text{bulk}}}\right), \tag{1}$$

where I_{SERS} is the intensity of enhanced Raman peak, I_{bulk} is the intensity of bulk RhB peak, C_{SERS} is the known concentration of RhB solution already and C_{bulk} is the concentration of bulk RhB at 2.75 mol/L. The result of EF is shown in Table 1, which illustrated that the highest enhancement factor for SERS is 2.47×10^5 in Ag sputtering for 30 s. This enhancement is shown at the position of 1674 nm because of the C=C aromatic stretching.[7]

The size-correlated enhancement can be explained by electromagnetic (EM) mechanism,[8,9] which explains that the Ag nanoparticles in the size

Table 1. The enhancement factor and Ag magnetron sputtering time.

Time	20 s	25 s	30 s
Enhancement factor	2×10^5	2.14×10^5	2.47×10^5

from 60 nm to 80 nm get the highest enhancement factor. SERS effect of Ag-TiO_{2-X} nanocomposite not only because of the EM mechanism, but also the TiO_{2-X} molecule CT mechanism. TiO_{2-X} could be excited by the laser; electrons are injected and accumulated at Ag nanoparticles, then transferred to the LUMO and HOMO orbit of the molecules.[10] In addition, TiO_{2-X} nanotubes contain defects in its crystal structure, which decrease the recombination of laser-excited electrons and holes.[11]

4. Conclusions

In this study, different size and density of Ag nanoparticles are deposited on the TiO_{2-X} nanostructured substrate by magnetron sputtering. SEM images show that the size and density of the Ag nanoparticles can be adjusted by varying the magnetron sputtering time, and a uniform Ag particle distribution has been obtained after dewetting treatment. The SERS enhancement of Raman spectrum showed an Ag size dependence on the Ag-TiO_{2-X} nanocomposite substrate. The best enhancement was obtained with a moderate Ag deposition time of 30 s. The relationships of metal nanoparticle size and SERS enhancement have been discussed, which can be extended to other metal particles such as Au, Pt and Cu.

Acknowledgments

T. Zhang would like to thank the support from New Zealand Government Scholarship.

References

1. D. M. Zhang *et al.*, *J. Mater. Sci.* **54**, 7576 (2019).
2. X. M. Zhou *et al.*, *Chem. Commun.* **55**, 533 (2019).
3. M. Plodinec *et al.*, *J. Alloy Compd.* **776**, 883 (2019).
4. H. R. Zhou *et al.*, *Nanomaterials* **8**, 12 (2018).
5. M. S. Zhu *et al.*, *Appl. Catal. B Environ.* **221**, 645 (2018).
6. C. Liu *et al.*, *Chem. Eng. J.* **359**, 58 (2019).
7. J. T. Zhang *et al.*, *J. Phys. Chem. B* **109**, 12544 (2005).

8. S. Y. Ding *et al.*, *Chem. Soc. Rev.* **46**, 4042 (2017).
9. S. J. Oldenburg, *Silver Nanoparticles: Properties and Applications* Available at: http://www.sigmaaldrich.com/materials-science/nanomaterials/silver-nanoparticles.html.
10. L. B. Yang *et al.*, *J. Phys. Chem. C* **113**, 16226 (2009).
11. X. D. Yan, Y. Li and T. Xia, *Int. J. Photoenergy* (2017).

https://doi.org/10.1142/9789819818433_0009

Chapter 9

Ultra Clean Manufacturing Techniques for Final Optics Assembly in High-Power Laser Facility[#]

Xiaodong Yuan[*,‡], Caizhen Yao, Xinxiang Miao, Haibing Lv, Guorui Zhou, Longfei Niu and Hao Liu

Laser Fusion Research Center, China Academy of Engineering Physics, Mianyang 621900, P. R. China

Feihu Zhang[†,‡], Qingshun Bai, Jiaxuan Chen, Yongzhi Cao and Yongbo Guo

Harbin Institute of Technology, Harbin 150001, P. R. China
[*]*xdyuan@caep.cn*
[†]*zhangfh@hit.edu.cn*

Cleanliness control is one of the most important processes for final optics assembly (FOA) in high-power laser systems. In this paper, an "Ultra Clean Manufacturing (UCM)" concept is put forward for the first time to satisfy the cleanliness requirements for FOA. First, the challenges and requirements for cleanliness in FOA are introduced and the concept of UCM is proposed. Significant surface cleaning technologies for mechanical parts and optics are then expounded. Finally, surface cleanliness control techniques, such as "gas knife" and "laser cleaning", are developed as online cleaning methods for the surface cleanliness in FOA.

Keywords: Ultra-clean manufacturing; final optics assembly; high-power laser facility; cleanness control; contamination.

1. Introduction

Comparing with other high-power lasers, the inertial confinement fusion (ICF) driver has three notable characteristics: large scale, high precision and

[‡]Corresponding authors.

[#]To cite this article, please refer to its earlier version published in the *International Journal of Modern Physics B*, Volume 34, 2040017 (2020), DOI: 10.1142/S0217979220400172.

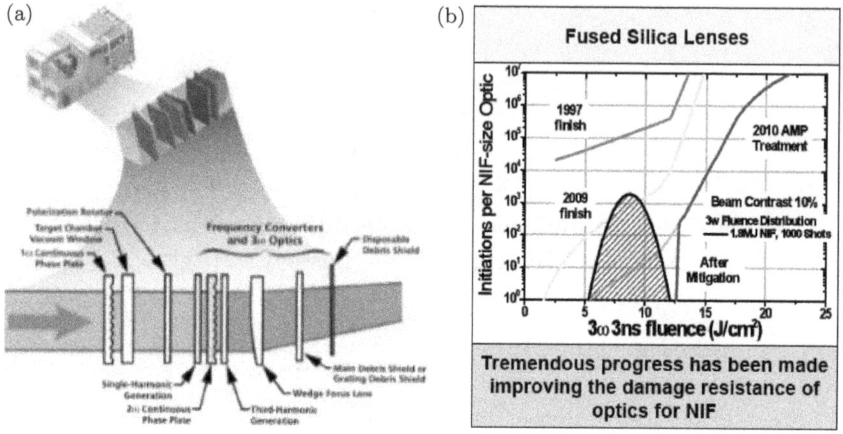

Fig. 1. (Color online) (a) The FOA of NIF[9] and (b) the damage resistance level of optics in NIF.[10]

super cleanliness.[1] The most vulnerable system in National Ignition Facility (NIF) is the final optics assembly (FOA), which is shown in Fig. 1(a). Scientists in Lawrence Livermore National Laboratory (LLNL) have made great efforts to increase the damage thresholds of the optics in FOA to improve the output ability of the laser system, and the damage threshold of fused silica in FOA [Fig. 1(b)] has been doubled or tripled by application of varies advanced manufacturing processes.[1–3] However, there are still fused silica components being damaged during operation. The main reason could be attributed to the surface contamination.[4–8] Preventing, controlling and eliminating the contaminants during laser shooting are critical to ensure the stable operation of a laser facility with high power.

2. The Concept of Ultra Clean Manufacturing

Ultra Clean Manufacturing (UCM) is defined as "The total science and technology to create and maintain the super-clean surface". According to this definition, UCM contains every dimension and every aspect of manufacturing. Figure 2(a) gives the dimensions of the UCM and Fig. 2(b) gives the surface cleanliness requirements for UCM. The definition of UCM surface satisfies the requirement of ISO SCP4 and IEST Class 25 according to ISO 14644-9 and IEST STD 1246C, and it was put forward for the first time in the world to satisfy the cleanliness requirement for high-power laser systems.

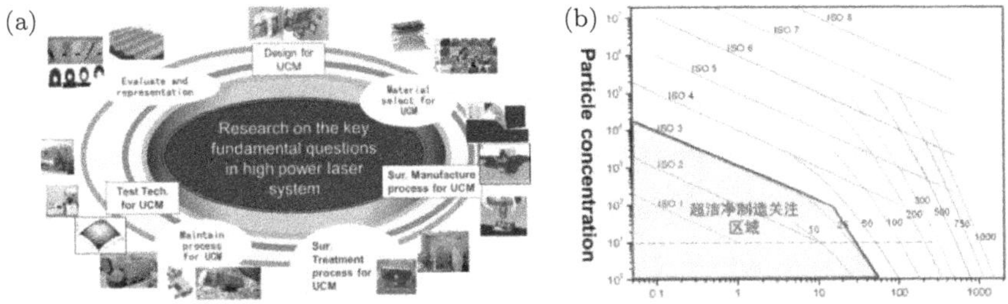

Fig. 2. (Color online) (a) Dimensions of UCM and (b) surface cleanliness requirement for UCM.

3. Key Technologies of UCM

3.1. *Surface cleaning technologies for mechanical parts*

3.1.1. *Precision wiping technology*

Precision wiping is widely used in surface cleaning. Organic and large particulates (>30 μm) can be removed efficiently. The most popular process is to wipe the surface by using organic solutions such as alcohol and isopropyl alcohol, and then heat the module at 80°C for several hours to further remove the molecular contaminants.

3.1.2. *High pressure water spray technology*

Figure 3 shows the high-pressure water spray technology which can eliminate organic compounds and particles larger than 5 μm on smooth surfaces. It has been widely used to clean the inner surfaces of large vessels and structures. The temperature, pressure, spray angle and the number of spraying times were investigated during the cleaning processes of stainless steels. Results indicate that the optimum temperature is 65°C, pressure is from 15 MPa to 17 MPa, and the maximum spray time is 4 min.

3.1.3. *Ultrasonic cleaning technology*

The primary physical phenomenon in ultrasonic cleaning is cavitation, which is the creation and subsequent collapse of microscopic bubbles in a liquid. These bubbles are formed when a large pressure gradient is introduced in a fluid. When the bubble grows, it can partially collapse or completely implode. These processes re-radiate the absorbed energy from the incident acoustic field. This re-radiated energy impinges upon

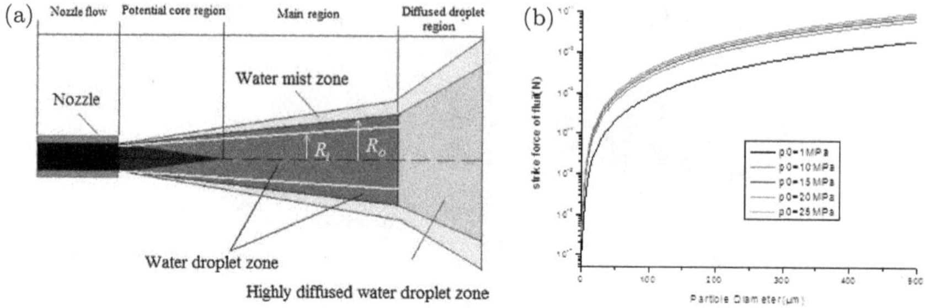

Fig. 3. (Color online) (a) The effective spray area and (b) cleaning forces for different particle diameters.

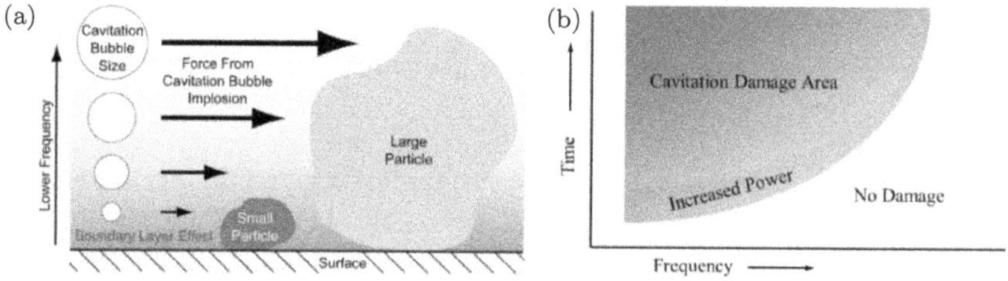

Fig. 4. (Color online) (a) The ultrasonic cleaning mechanism and (b) cavitation damage versus frequency.

a substrate and does the bulk cleaning. The cleaning efficiency largely depends on the frequency and power of cleaner. In this research, 28 kHz was applied for rough cleaning, and 40 kHz, 108 kHz were used for precision cleaning. The ultrasonic cleaning mechanism is shown in Fig. 4.

3.2. *Surface cleaning technologies for optics*

3.2.1. *Ultrasonic and mega-sonic cleaning technologies for fused silica optics*

Fused silica optics are mainly used as the final optics, and they have the highest damage risk in the laser facility. A perfectly fused silica surface is needed for the high fluent operation. An acid etching associated with mega-sonic wave combined with ultrasonic cleaner [Fig. 5(a)] has been developed to get a perfect surface of fused silica optics.[11,12] During the etching process, the mega-sonic wave will take the residue of the chemical reaction products away from the surface. After acid etching process, the acid will be retrieved

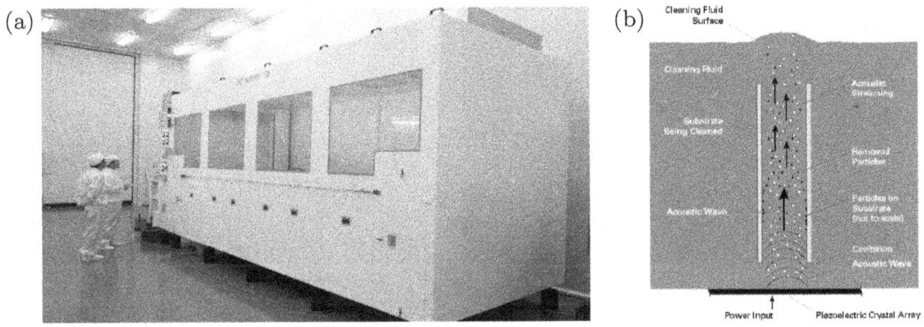

Fig. 5. (Color online) (a) Ultrasonic and mega-sonic cleaner for fused silica optics and (b) cleaning principle.

and the fused silica elements will be transferred into the second tank, in which high-pressure pure water spray is used to remove the residual HF acid from optics. The maximum pressure is 0.5 MPa, and highest temperature is 65°C. Then, ultra-sonic cleaners are used in the third and the fourth tanks, in which complex multi-frequency transducer is used to generate multi-frequencies, including 40, 80, 120, 170, 210, 240 and 270 kHz, to remove particulates larger than 0.5 μm. At last, a slow dip process is used to dewater the elements.

3.2.2. *The precision cleaning technologies for dielectric film coated optics*

The dielectric film coated optics are widely used to transmit laser pulse, they cannot be washed by water-based solution. The traditional cleaning process is to wipe the surface by organic solutions, such as alcohol and isopropyl alcohol, which is difficult to satisfy for the high-level cleanliness requirements.[13] According to the nature of dielectric film coated elements, laser cleaning technology is developed to obtain high surface cleanliness. Results show that the laser cleaning technology can remove small particles. For large particles, it may damage the coating during the cleaning process. The images of a dielectric film coated optic before and after cleaning are shown in Fig. 6.

3.2.3. *The precision cleaning technologies for neodymium glasses*

The neodymium glasses are used as the gain medium to amplify the laser energy. Compared with fused silica optics, the surface of neodymium glass

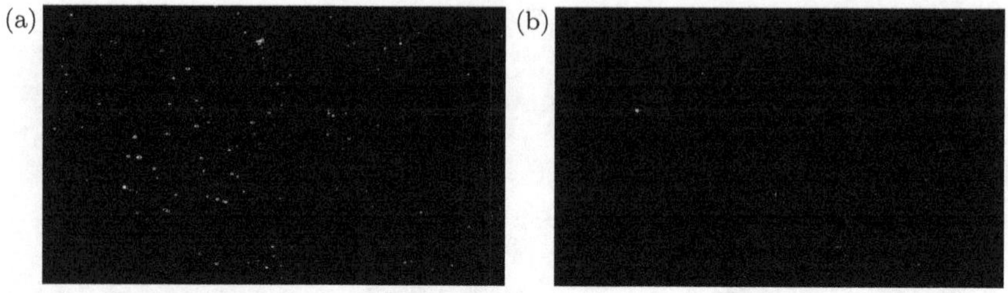

Fig. 6. The cleaning result and damage of a dielectric film coated optic: (a) the dark field image before cleaning and (b) the dark field image after cleaning.

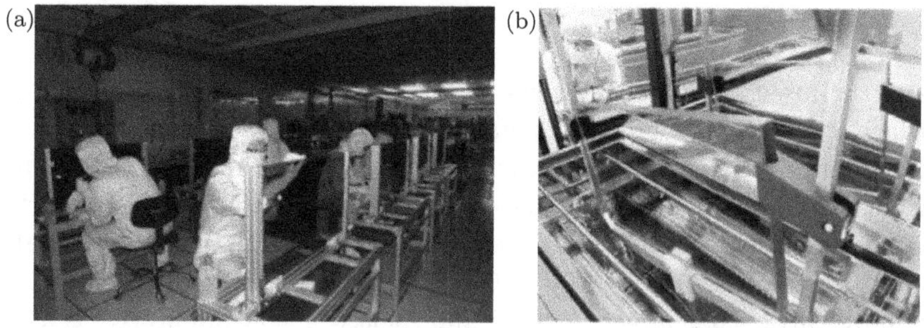

Fig. 7. (Color online) The washing process for neodymium glass: (a) cleaning the surface by precision wiping and (b) ultra-sonic cleaning by using pure water.

is soft, readily weathered and easy to deliquesce and be scratched. The traditional cleaning method is to wipe the surface with water-free organic solvent, which is time-consuming and not efficient for high cleanliness requirement. Thus, a new method is developed by using pure water (15 MΩ). The warm pure water (45°C) is sprayed onto the surface with high pressure (0.5–0.7 MPa) for 1 min followed by ultra-sonic cleaning for 3 min. At last, a water-free solvent is used to dry the surface immediately. The cleanliness by this method is better than ISO SCP 4 without using any deliquesces. The result is shown in Fig. 7.

3.3. *Surface cleanliness control technologies*

3.3.1. *Gas knife sweep technology*

Large particles on an optical surface can be removed by using a gas knife. According to the experimental results, more than 95% particles that are

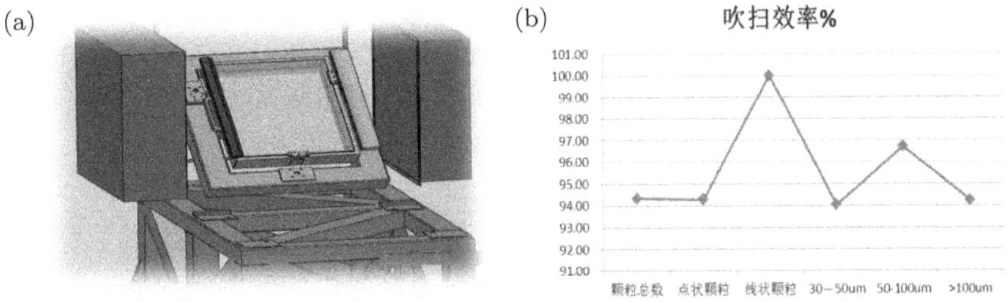

Fig. 8. (Color online) (a) The design of the gas knife and (b) sweep efficiency.

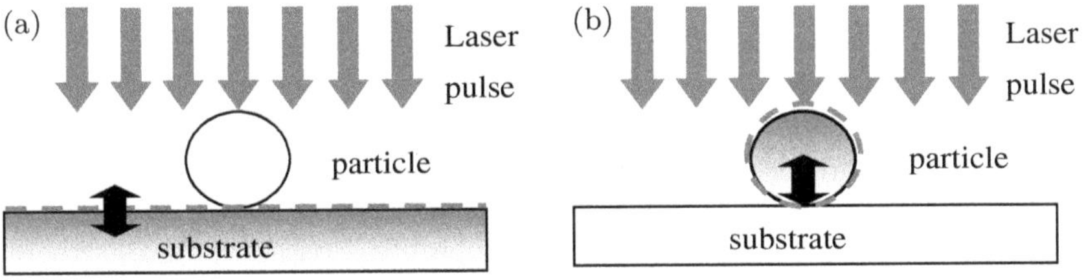

Fig. 9. (Color online) A schematic representation of dry laser cleaning process: (a) strong substrate absorption and (b) strong particle absorption.

larger than 30 μm can be removed when the gas pressure is from 0.6 MPa to 0.8 MPa with an output gas speed higher than 70 m/s. A static electric absorber is used to avoid the re-deposition of particles. Figure 8 shows the design of the gas knife and sweep efficiency.

3.3.2. *Laser cleaning technology*

The principle of dry laser cleaning is shown in Fig. 9. When laser pulse irradiates on the surface, contaminants or the substrate will absorb the laser energy and generate a high-temperature within extremely short time duration ($\sim 10^{-8}$ s). Particulate contaminants will obtain a high acceleration velocity (10^9 m/s^2) by thermal expansion.[1,2] They will then overcome the van der Waals force and jump up from the surface.

4. Conclusions

The concept of the UCM was proposed, which is to take cleanliness into consideration during the whole manufacturing process. These methods were put forward and verified in the laser facility, with examples of surface cleanliness control of the mechanical parts and optics in our laser facility. The progress of UCM for high-power solid-state laser system is systematically explained, which established the foundation for the next generation of large high-power solid-state laser facilities.

Acknowledgments

This work is supported by the National Natural Science Foundation of China (Nos. 51535003 and 51701087), the Special foundation for Young Scientists of LFRC (Grant No. RCFPD3-2019-3) and the Special foundation for Young Scientists of CAEP (Grant No. PY2019109).

References

1. Y. Huag *et al.*, *Comput. Phys. Commun.* **185**, 459 (2014).
2. G. T. Wirtenson, *Opt. Eng.* **18**, 574 (1979).
3. J. Y. Natoli, B. Bertussi and M. Commandre, *Opt. Lett.* **30**, 1315 (2006).
4. X. X. Miao *et al.*, *High Power Laser and Particle Beams* **4**, 890 (2013).
5. C. Macilwain, *Nature* **401**, 201 (1999).
6. X. X. Miao *et al.*, *High Power Laser and Particle Beams* **22**, 1740 (2010).
7. X. X. Miao *et al.*, *Chin. J. Lasers* **21**, 9 (2015).
8. S. R. Qiu *et al.*, *Appl. Opt.* **54**, 8607 (2015).
9. National Ignition Facility User Guide, Lawrence Livermore National Laboratory (2012), p. 48.
10. National Ignition Campaign Program Completion Report, National Security Administration, 2012, 120.
11. L. Yang, *Chin. Phys. B* **25**, 632 (2016).
12. H. Wang *et al.*, *High Power Laser and Particle Beams* **27**, 123 (2015).
13. L. Yang *et al.*, *Opt. Eng.* **54**, 126101 (2015).

https://doi.org/10.1142/9789819818433_0010

Chapter 10

Investigation of New Zealand's Natural Magnetic Minerals for Application in Inroad Charging Systems[#]

William J. Trompetter[*,§], Jérôme Leveneur[*], John Kennedy[*], Ben Rumsey[†], Murray McCurdy[†], Shen Chong[‡] and Nick Long[‡]

[*] *National Isotope Centre, GNS Science, 30 Gracefield Road, Lower Hutt, New Zealand*

[†] *CRL Energy, 68 Gracefield Road, Lower Hutt, New Zealand*

[‡] *Robinson Research Institute, Victoria University of Wellington, 69 Gracefield Road, Lower Hutt, New Zealand*

[§] *b.trompetter@gns.cri.nz*

There has been a steadily increasing demand for magnetic materials in devices and charging systems over recent decades. The advent of in-road inductive power transfer (IPT) charging systems in the near future will massively increase the demand. Better, nonbrittle and affordable magnetic materials are required so that vehicles can run over charging pads built into the road without destroying them. This will require innovative solutions and new magnetic material sources to meet these needs in an economically viable way. Here, we have scoped New Zealand's (NZ) natural magnetic materials and their ability to be included as a viable magnetic material for our roads. We have investigated the magnetic permeabilities of NZ natural magnetic mineral deposits for IPT application in roads.

Keywords: Magnetic materials; magnetic minerals; charging systems; inductive power transfer.

1. Introduction

Recently, inductive power transfer (IPT) has been proposed as a possible tool to increase the usage of electric vehicles (EVs) by improving

[§]Corresponding author.

[#]To cite this article, please refer to its earlier version published in the *International Journal of Modern Physics B*, Volume 34, 2040018 (2020), DOI: 10.1142/S0217979220400184.

user-friendliness — removing the hassle of cables and mechanical connections, and increasing their practical range — with widely spatially distributed charging. In New Zealand (NZ), the Ministry of Transport's (MoT) latest electric vehicle projections (2017)[1] expect EV fleet numbers to at least double every year to reach 64,000 EVs by the end of 2021 and will make up 40% of the total vehicle fleet by June 2040. This estimate is in line with global estimates, with Bloomberg expecting EVs to account for 54% of sales by 2040.[1,2] Infrastructure to support this growth, or accelerating the technology's acceptance, is not deployed as yet.

Current IPT systems typically use ferrite materials to improve magnetic coupling, which are particularly suited for inductive applications operating between 10 kHz to 500 kHz.[3,4] Although ferrite is optimal for guiding magnetic fields, it is very brittle and would not survive within a road structure. Charging pads require nonbrittle and affordable magnetic materials which retain their properties when subjected to the stresses of an active roadway. This requires innovative solutions and new magnetic material sources to meet these needs in an economically viable way.

A potential way is using soft magnetic composite (SMC) materials in which a magnetic filler is mixed into a matrix to optimize economic viability and the material's mechanical, thermal and magnetic properties. SMC materials have a critical role in power electronics, enabling high performance inductors, transformers, inverters, and other inductive devices.[5] These materials have been extensively studied over the past 50 years resulting in a wide range of materials with properties optimized for given applications. A German company, Magment, has recently developed a magnetic concrete (MC40) consisting of magnetic particles with sizes ranging from around hundreds of μm to several mm mixed in a cement matrix. Similarly, polymer SMC materials can be found commercially with a range of magnetic properties and are potentially less brittle. However, their cost is currently significantly higher than the aforementioned materials.[6]

A lower-cost magnetic material can be found in iron sand. This investigation considers the magnetic properties of NZ iron sand deposits and their provenance as a first step to assess potential usability in this application.

2. NZ Magnetic Mineral Resource

The magnetic properties of rocks are due almost entirely to their constituent iron-bearing minerals, particularly ferrimagnetic minerals such as (titano) magnetite, hematite, ilmenite and pyrrhotite. Magnetic properties depend on one or more of the composition, concentration, and grain-size distribution of the ferrimagnetic minerals. The main type of iron ore deposits in NZ are titanomagnetite and ilmenite rich coastal sands (iron sands).[7,8] Titanomagnetite iron sand beach and dune deposits on the west coast of the North Island, New Zealand, are some of the largest deposits of this type in the world and are described in the NZPAM report "Mineral Commodity Report 15 — Iron"[8] and references therein. The titanomagnetite iron sand deposits of the west coast, North Island, contain a total identified resource of over 850 Mt of concentrate assaying 55–56% Fe, 7–9% TiO_2 and 0.3–0.4% $V_2O_5^8$.[8] Similar magnetite bearing coastal sand deposits in Japan, the Philippines and Indonesia have been mined for iron ore to a limited extent. Recently,[9] it was found there were two source regions: the northern region (Taharoa northwards) composed primarily of titanomagnetite present in the rivers that drain the Taranaki volcanoes, and the southern sector consistent with titanomagnetite from Mt. Taranaki andesites.

3. Materials and Methods

For this investigation, iron sand samples from locations around NZ were obtained by Coal Research Limited (CRL). Their magnetic component was magnetically separated from sand samples with a Davis tube set and with a magnetic field of 1500 Gauss. The weight of the magnetic component/initial sample = magnetic fraction (% mag.). Table 1 lists the sample locations and the percentage of magnetic minerals is presented in Fig. 1. The Taharoa and Waikato North Head were sourced from working mines-feed, concentrate or tailings. The Barrytown sample was a composite from exploratory drill cores. The remaining beach and dune samples were small surface grab samples from areas obviously higher in magnetics (visually and rough magnet test). The samples' composition was determined using Ion Beam Analysis (IBA).[10] Magnetization saturation (M_s) in emu/g and saturation field in H_s were obtained using vibrating sample magnetometry (VSM) at 298 K. VSM measurements were made with a sweeping field rate of 100 Oe/min upto ±1 T magnetic field. The samples were held in a gelatine capsule and weighed prior to measurements.

To measure the AC magnetic properties, the iron sand samples were loaded into toroid-shaped 3D printed containers. The samples were weighed, and their density calculated showing a dispersion of 11% (1 standard deviation). The dimensions of the toroid containers (as shown in Fig. 2) were: height $h = 9.0$ mm, inner radius $a = 4.5$ mm and outer radius $b = 8.0$ mm. The toroids were wound with insulated copper wire ($N = 10$ turns) and

Table 1. Results summary (Lat. = latitude, Long. = longitude).

#	Lat.	Long.	Location	% mag	μ_r	M_s [emu/g]	H_s [Oe]	Mg [ppm]	Al [ppm]	Si [ppm]	K [ppm]	Ca [ppm]	Ti [ppm]	Cr [ppm]	Fe [ppm]	Co [ppm]	O [ppm]
1	−37.35	174.74	Waikato North Head	56	4.85	50.2	2564	7140	20,172	20,947	159	1831	39,535	143	665,535	2086	366,587
2	−38.16	174.72	Taharoa	35	3.65	46.0	3071	8900	15,759	17,953	564	4166	41,678	83	650,094	1687	376,033
3	−38.68	174.62	Awakino Dunes	94	4.23	47.1	2709	19,503	38,282	84,558	2335	8968	39,946	31	654,383	3001	364,432
4	−38.68	174.62	Awakino Beach	75	4.05	48.7	2674	17,836	37,364	64,997	1753	8914	41,973	125	698,656	2191	369,060
5	−38.70	174.62	Mokau Beach	73	4.37	43.2	2440	17,326	27,485	42,308	829	8460	40,300	310	689,477	3540	363,606
6	−38.99	174.22	Waitara Beach	80	3.10	40.3	2545	28,281	45,179	102,829	2353	17,874	36,551	160	582,894	1866	388,134
7	−38.99	174.22	Waitara Dune	45	2.41	33.9	2620	31,508	69,895	177,669	5156	23,575	28,043	27	426,002	929	383,877
8	−39.03	174.11	Te Rewa Rewa Beach	38	1.53	25.4	2508	52,652	83,556	245,620	6289	41,130	23,258	78	372,484	265	417,449
9	−39.05	174.10	Fitzroy Beach	24	1.15	19.7	2591	69,082	110,777	349,296	10,281	48,106	19,409	85	323,222	942	454,254
10	−39.61	174.24	Waihi Beach Hawera	63	4.04	47.2	2227	12,855	23,528	33,237	588	6791	32,486	195	608,701	2650	342,309
11	−39.77	174.49	Patea Sand dune	60	4.28	43.2	2390	26,417	30,545	81,586	868	20,412	37,181	151	621,164	3012	379,926
12	−39.77	174.49	Patea Beach	55	4.67	44.0	2423	15,550	25,160	37,602	584	8108	39,377	180	687,449	2,774	357,239
13	−39.83	174.64	Waverley Sand Dune	26	2.78	34.4	2239	21,187	38,176	83,001	1444	14,362	37,722	120	630,318	2792	379,672
14	−39.83	174.64	Waverley Beach	53	4.07	42.5	2391	15,539	28,004	41,379	865	8179	38,679	95	662,710	2,759	356,521
15	−39.88	174.90	Kai Iwi River Mouth	38	3.35	36.2	2382	17,548	38,482	74,769	1661	13,236	40,328	139	671,114	3207	387,811
16	−40.90	176.22	Castle Point Beach	8.9	3.98	44.8	2685	30,220	36,617	89,514	1538	15,736	33,943	126	560,993	1448	350,712
17	−42.22	171.32	Barrytown	0.2	0.69	1.9	8215	19,431	34,569	60,109	981	7013	37,442	116	600,871	727	431,708

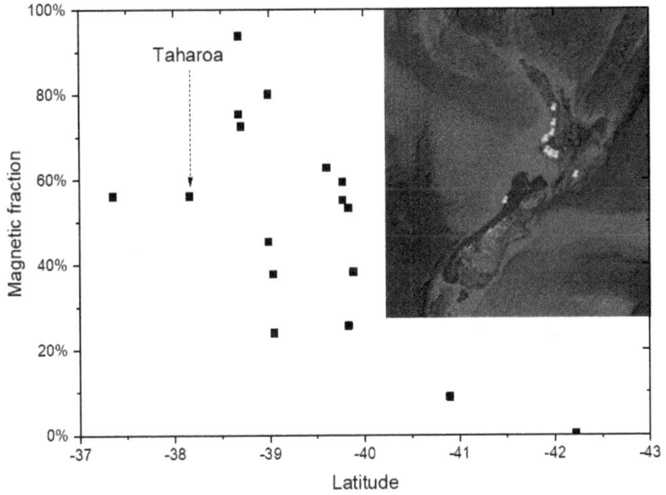

Fig. 1. (Color online) Iron sand sample locations and their magnetic fraction.

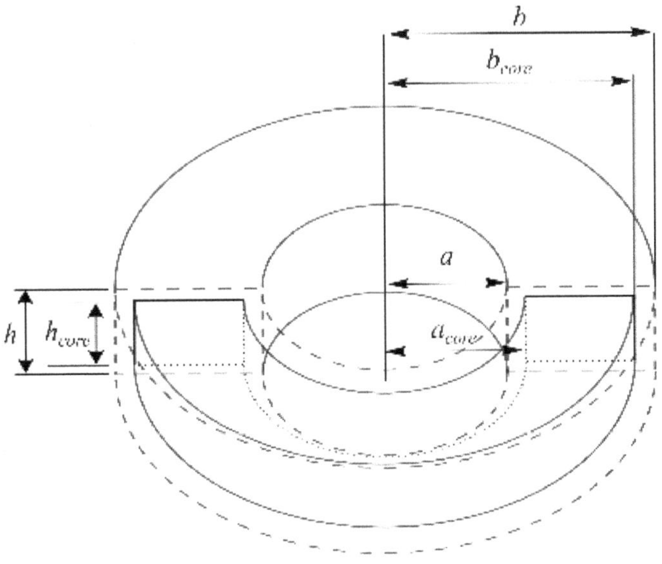

Fig. 2. Geometry of toroid sample and container.

connected to an LCR meter (Motech model MT4080A). An LCR meter measures the inductance (L), capacitance (C) and resistance (R) of the samples. The LCR meter makes the measurements at preset frequencies of 100 Hz, 120 Hz, 1 kHz, 10 kHz and 100 kHz. As the target AC frequency for inroad charging systems is currently 85 kHz, we report the results of the 100 kHz measurements in this study.

The inductance of a toroid in the absence of a magnetic core ("air core") is given by $L_0 = \frac{\mu_0 N^2 h}{2\pi} \ln \frac{b}{a}$ and the inductance with a sample of relative magnetic permeability, μ_r, is $L_{\text{sample}} = \frac{\mu_0 \mu_r N^2 h}{2\pi} \ln \frac{b}{a}$. When the dimensions of the core (a_{core}, b_{core} and h_{core}) are different to that of the toroid encased by the coil, such as in the case of our samples, the permeability is found by subtracting L_0 to the inductance measured with the core as shown in Eq. (1).

$$\mu_r = 1 + \left((L_{\text{sample}} - L_0) \frac{2\pi}{\mu_0 N^2 h_{\text{core}} \ln \frac{b_{\text{core}}}{a_{\text{core}}}} \right). \tag{1}$$

4. Results and Discussion

The magnetic properties of the natural magnetic mineral samples from around NZ obtained by CRL Energy Limited are displayed in Table 1.

Figure 1 shows that the magnetic fraction generally decreases toward the south, while the two most northern sites are around 60% magnetic which is consistent with published distributions.[9]

The magnetization saturation and relative magnetic permeabilities of the magnetic fraction of these samples do not show a significant relationship with latitude. However, these magnetic properties were affected by the composition of the samples. Figure 3 shows that higher permeabilities were observed with higher magnetic element concentrations of Fe (Co also has a similar relationship), conversely higher permeabilities occurred with less crustal element concentrations (Si + Al + Mg + K + Ca). The oxygen content also has a negative correlation with magnetic permeability corresponding to the higher oxygen content of crustal minerals relative to magnetite minerals.

In Table 2, the magnetic permeability of titanomagnetite concentrate is compared with commercial IPT materials. While SMCs have lower magnetic permeability (40–80 versus ~2,300 at 100 kHz) and lower saturation than CF295 ferrite (< 300 mT versus 525 mT), they offer flexibility to find an optimum set of electromagnetic, mechanical and thermal properties. NZ magnetic minerals outlined in this paper offer a magnetic permeability of

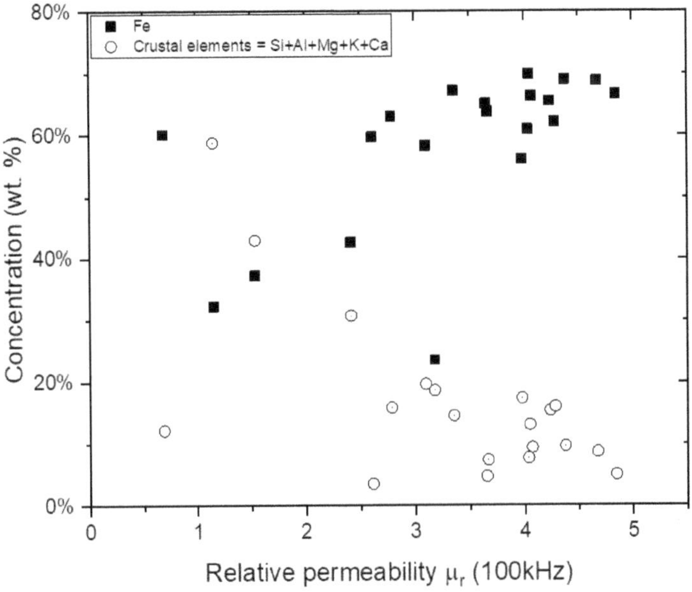

Fig. 3. Relative magnetic permeability versus iron and crustal element content.

Table 2. Comparison of titanomagnetite concentrate with commercial IPT materials.

No.	IPT materials	Manufacturer Specification μ_r @ 100 kHz	Effective Permeability μ_r @ 100 kHz	Density g cm^{-3}	Thermal Conductivity W/m K
1	Current IPT material: Ferrite Mn-Zn (CF295)-solid block	$3000 \pm 20\%$	2300	4.9	5
2	Current IPT material: Ferrite Mn-Zn (CF295)-crushed	NA	30	4.9	5
3	Commercial SMC-Fluxtrol 100	80	74	6.8	23
4	Commercial SMC-Fluxtrol 50	36	51	6.1	6
5	Commercial SMC-magnetic concrete (MC40)	40	44–54	3.8	3
6	NZ Titanomagnetite mineral-concentrate	NA	1–5	4.8	NA

1–5 with a fill fraction of ∼80% which is at the lower end of the commercial SMC materials, however, they could be combined with other materials to provide a viable affordable magnetic material robust enough for inroad charging systems.

5. Summary

Current IPT systems typically use ferrite materials due to their excellent high-frequency magnetic permeability characteristics, however, ferrite is brittle and would not survive within a road structure. Soft magnetic composites (SMC) comprising electrically insulated ferromagnetic particles are typically less brittle and have the added advantage of having low magnetic losses in high frequency applications.

NZ magnetic minerals outlined in this paper offer a magnetic permeability of 1–5, which is at the lower end of the commercial SMC materials. However, they could be combined with other materials to provide a viable and affordable magnetic material robust enough for inroad charging systems. The viability of using iron sand as an SMC composite would be enhanced if further optimization is undertaken. The elemental analysis revealed that higher magnetic performance of the natural iron sand magnetic minerals can be achieved through the selection of magnetic minerals with lower crustal elements or by the improved removal of the crustal fraction. Further analyses will be undertaken to understand the difference in magnetic properties of the different titanomagnetite minerals.

Acknowledgment

We acknowledge Taharoa Ironsands Ltd. for providing bulk iron sand samples and Trevor Dine from CRL Ltd. for beach and dune sample collection.

References

1. Ministry of Transport, Electric Vehicles Programme, http://www.transport.govt.nz/ourwork/climatechange/electric-vehicles/.
2. https://about.bnef.com/electric-vehicle-outlook/#toc-download.
3. G. A. Covic and J. T. Boys, *IEEE J. Emerging Sel. Top. Power Electron.* **1**, 28 (2013).
4. B. Park *et al.*, *Int. J. Commun.* **2**, 59 (2017).
5. R. C. O'handley, *Mod. Magn. Mater.* (2000).
6. https://fluxtrol.com/fluxtrol-50.

7. Officers of the New Zealand Geological Survey: "Minerals of New Zealand (Part A: Metallics, 2nd edn.)", *New Zealand Geological Survey Report* **38A**, (1981).
8. A. B. Christie and R. L. Brathwaite, NZPAM report "Mineral Commodity Report 15 — Iron" (1997), https://www.nzpam.govt.nz/assets/Uploads/doing-business/mineral-potential/iron.pdf.
9. R. L. Brathwaite, M. F. Gazley and A. B. Christie, *NZ J. Geochem. Explor.* **178**, 23 (2017).
10. W. J. Trompetter, A. Markwitz and P. K. Davy, *Int. J. PIXE* **15**, 249 (2005).

https://doi.org/10.1142/9789819818433_0011

Chapter 11

Cu_2ZnSnS_4 (CZTS) Thin Films Prepared by Sol–Gel Spin-Coating Technique[#]

Chao Xiong[*,†,‡], Min Gao[*] and Wei Gao[†]

[*]*School of Photoelectronic Engineering, Changzhou Institute of Technology, Changzhou 213032, P. R. China*

[†]*Department of Chemical & Materials Engineering, The University of Auckland, PB 92019, Auckland 1142, New Zealand*

[‡]*xiongc@czu.cn*

The Cu_2ZnSnS_4 (CZTS) thin film is a photovoltaics material with excellent photoelectric properties and has good potential applications. The CZTS thin films with different spin-coated layers and sulfurized temperatures were successfully prepared by a sol–gel spin-coating technique. The relationships of microstructure, surface morphology and processing parameters were studied using XRD, SEM and EDS. The results indicate that the grain size of the sample sulfurized at 600°C was larger than that of the sample sulfurized at 550°C. With increasing layers of spin, coatings, the films showed better crystalline structure. The sulfureted CZTS thin films prepared by six spin-coating possess uniform elemental distribution.

Keywords: Cu_2ZnSnS_4; sol–gel method; microstructure; photoelectric properties.

1. Introduction

Cu_2ZnSnS_4 (CZTS) has high absorption efficiency (~10^4/cm) and suitable optical band structure of about 1.5 eV which is considered as the promising absorber for low-cost thin film solar cells.[1–5] In recent years,

[‡]Corresponding author.

[#]To cite this article, please refer to its earlier version published in the *International Journal of Modern Physics B*, Volume 34, 2040019 (2020), DOI: 10.1142/S0217979220400196.

the development of CZTS thin film solar cells has become one of the hot research topics.[6–8] CZTS thin film solar cells were prepared for the first time in 1996, and their efficiency was only 0.66%.[9] The conversion efficiency was improved slowly till 2006, the research on CZTS increased rapidly, and the efficiency of the cell was constantly refreshed. At present, the efficiency of CZTS has exceeded 12%.[10] CZTS thin film solar cells could be fabricated by various methods and different materials synthesis methods. The device efficiency and performance can be very different.[11–14] Developing low-cost methods has always been one of the research directions of CZTS thin films.

In this work, the CZTS thin film was successfully prepared by a sol–gel spin- coating method. The microstructure and surface morphology of CZTS thin films with different number of spin-coated layer and sulfurized temperatures were studied. The crystalline quality of the samples was measured by X-ray diffraction (XRD). The surface morphology of the samples was examined by scanning electron microscopy (SEM) and energy dispersive spectrometer (EDS) mapping.

2. Experimental

2.1. *Precursor sol solution*

The precursor solution was prepared by dissolving $Cu(CH_3COO)_2 \cdot H_2O$ (1.278 g, AR), $Zn(CH_3COO)_2 \cdot 2H_2O$ (1.098 g, AR), $SnCl_2 \cdot 2H_2O$ (0.903 g, AR) and $SC(NH_2)_2$ (2.436 g, AR) into 2methoxyethanol (20 ml, AR) while stirring at 60°C for 4 h to get clear and transparent yellow solution. After aging at room-temperature air for three days, the solution has a certain viscosity, which could be used for subsequent spin, coating. The process for preparation of precursor Sol solution is shown in Table 1. All chemical reagents were purchased from Wuxi Yatai Chemical Co.

Table 1. Process for preparation of precursor sol solution.

Symbol	Chemical substances	Quantity	First step	Second step	Third step
A	$Cu(CH_3COO)_2 \cdot H_2O$	1.278 g	A, B, C and D are added to E in turn and stirred continuously	Heating at 60°C and fully stirring for 4 hours on a magnetic stirrer	After aging for 3 days
B	$Zn(CH_3COO)_2 \cdot 2H_2O$	1.098 g			
C	$SnCl_2 \cdot 2H_2O$	0.903 g			
D	$SC(NH_2)_2$	2.436 g			
E	2methoxyethanol	20.0 ml			

2.2. *CZTS thin films preparation*

The prepared precursor sol solution was spin coated on ITO glass substrates. First, it spun at 500 rpm for 30 s, and then increased to 1500 rpm for 45 s. The next step was to heat the spin-coated sample in a furnace at 200°C for 6 min in air to remove moisture and organic matter. The spinning steps of the coating were repeated 3, 6 and 9 times to obtain CZTS thin film precursors with different thickness. Afterward, the as-prepared CZTS thin film precursors were sulfurized in a tube furnace with appropriate amount of sulfur powder at 550°C and 600°C for 20 min in N_2 atmosphere, forming the sulfurized CZTS thin films.

2.3. *Characterization and analysis*

The surface morphology of thin films was characterized by a field emission scanning electron microscope (Zeiss-Sigma). The XRD patterns were collected by using X-ray diffractometer (Rigaku Ultima IV). EDS, Oxford instruments, was used to observe the chemical composition of the composites.

3. Results and Discussion

3.1. *Microstructure of coatings*

The XRD patterns of CZTS thin film with 3, 6 and 9 different number of spin-coatings are shown in Fig. 1. All the CZTS thin films have a crystalline structure. The three main diffraction peaks correspond to (112), (220) and (312) crystal planes of CZTS structure at 28.5°, 47.3° and 56.1°, respectively. The intensity of the diffraction peaks on (112) plane is the highest. The patterns of the samples are matched with the kesterite CZTS (JCPDS No. 26-0575). The (200) plane of In_2O_3 in ITO glass substrates is the diffraction peak appearing at about 35°C. According to the width and height of the diffraction peak on (112) plane, the grain size of the sample sulfurized at 600°C is larger than that of the sample sulfurized at 550°C. This is because the higher the annealing temperature is, the easier the grain growth is. It can also be seen from Fig. 1 that with the increasing layers of spin-coatings, the film has better crystalline structure. This may be because the thinner the film, the lower the grain density of CZTS, which is not favorable to the crystallization process.

To further verify the composition of the samples, EDS mapping was performed. The elemental distributions of Cu, Zn, Sn and S are shown

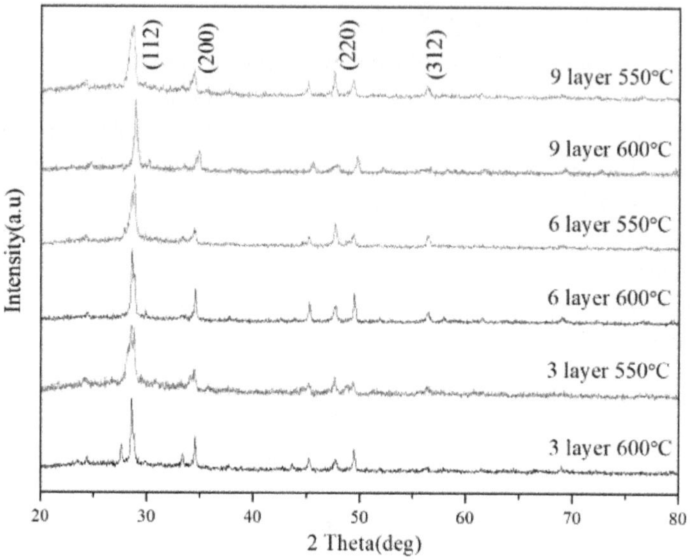

Fig. 1. (Color online) XRD patterns of CZTS thin films with 3, 6 and 9 spin-coated layers.

in Figs. 2(a)–2(d), respectively. As can be seen in Fig. 2, the distribution of each element is relatively uniform. The ratio of Cu, Zn, Sn and S in the sulfurized CZTS thin film spin-coated for six times is 4.8 : 2.6 : 1.8 : 10.4, which is close to the standard CZTS 2 : 1 : 1 : 4.

3.2. *Surface of coatings*

Figure 3 illustrates the surface morphologies of CZTS thin films with different number of spin-coated layers and different sulfurized temperatures. The surfaces of CZTS thin films with different spin-coated layers were rather smooth and homogeneous. Figures 3(a), 3(b), 3(d) and 3(e) show that the surface grains of CZTS films prepared with three layers are not dense and the grain size is smaller than those prepared by spin-coated six layers at the same sulfurized temperature of 500°C. This is consistent with XRD test results. Comparing Figs. 3(b), 3(c), 3(e) and 3(f), it can be seen that the grain size of CZTS films sulfurized at 600°C is larger than that of films cured at 550°C.

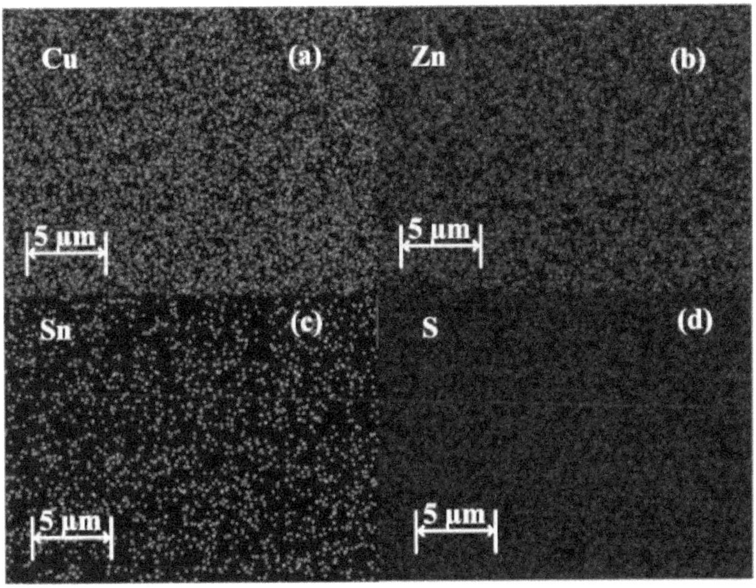

Fig. 2. (Color online) EDS mapping images of Cu, Zn, Sn and S in CZTS thin films.

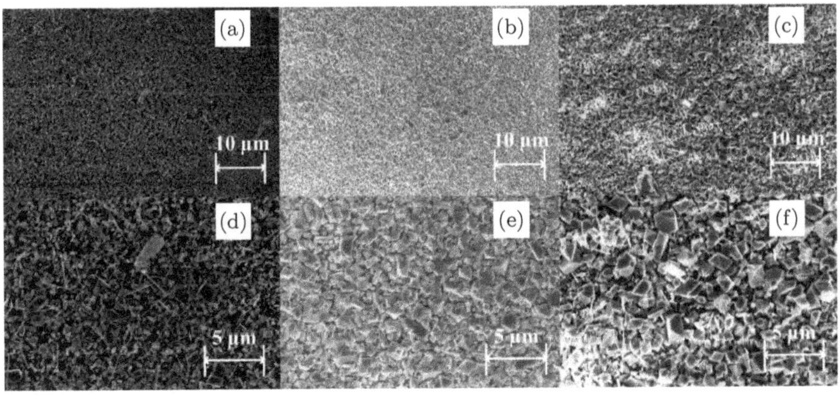

Fig. 3. The surface morphology of CZTS thin films with different spin-coated layers and sulfurized temperatures: (a) and (d) CZTS films spin-coated for three times and sulfurized at 550°C; (b) and (e) CZTS thin films spin-coated for six times and sulfurized at 550°C; (c) and (f) CZTS thin films spin-coated for nine times and sulfurized at 600°C.

4. Conclusions

The structure and surface morphology of CZTS thin films prepared by different spin-coating layers and annealing temperatures were investigated. The three main diffraction peaks correspond to (112), (220) and (312) crystal planes of CZTS structure at the positions of 28.5°, 47.3° and 56.1°, respectively. The grain size of the sample sulfurized at 600°C was larger than that of the sample sulfurized at 550°C. With the increasing spin-coated layers, the films show better crystallinity. The sulfureted CZTS thin films prepared by 6 spin-coated layers possess uniform elemental distributions. The ratios of Cu, Zn, Sn and S were 4.8 : 2.6 : 1.8 : 10.4, which are close to the standard stoichiometry of CZTS 2 : 1 : 1 : 4.

Acknowledgments

This project is supported by the Industry-University-Research Cooperation of Jiangsu Province (No. BY2018140), the Qing Lan Project of Jiangsu Higher Education, the Science and Technology Program of Changzhou (Nos. CE20180041, CE20180052 and CZ20180015), and the Key Laboratory of Electronic Measurement Technology, Anhui Province (The 41_{st} Institute of CETC, Grant No. YFKM-WHH-201705-01).

References

1. J. Li, R. Yao and C. Xiong, *Mster. Lett.* **201**, 97 (2017).
2. T. K. Todorov *et al.*, *Adv. Energy Mater.* **3**, 34 (2013).
3. R. Mainz *et al.*, *Nat. Commun.* **5**, 3133 (2014).
4. S. A. Vanalakar *et al.*, *J. Alloy. Compd.* **619**, 109 (2015).
5. S. Ahmed *et al.*, *Adv. Energy Mater.* **2**, 253 (2012).
6. K. Yu and E. A. Carter, *Chem. Mater.* **27**, 2920 (2015).
7. F. Liu *et al.*, *Appl. Mater. Interfaces.* **7**, 14376 (2015).
8. A. D. Collord and H. W. Hillhouse, *Chem. Mater.* **27**, 1855 (2015).
9. H. Katagiri *et al.*, *Appl. Phys. Express* **1**, 041201 (2008).
10. W. Wang *et al.*, *Adv. Energy Mater.* **4**, 1301465 (2014).
11. J. Li *et al.*, *Mater. Lett.* **210**, 20 (2018).
12. R. Touati, M. B. Rabeh and M. Kanzari, *Thin Solid Films.* **582**, 198 (2015).
13. D. B. Mitzi, *Adv. Mater.* **21**, 3141 (2009).
14. C. Yan *et al.*, *Nat. Energy* **3**, 764 (2018).

https://doi.org/10.1142/9789819818433_0012

Chapter 12

Model-Based Analysis of Geometrical Effects in Microfluidic Fuel Cell with Flow-Through Porous Electrodes#

Qiang Xu*, Yiyi She† and Li Li‡,§

*Department of Computer Science,
Changzhou Institute of Technology,
Changzhou 213032, P. R. China

†School of Energy and Environment,
City University of Hong Kong,
Hong Kong, P. R. China

‡School of Automotive and Traffic Engineering,
Jiangsu University of Technology,
Changzhou 213001, P. R. China
§liliorigin@126.com

Porous electrodes in microfluidic fuel cell (MFC) operate with nonuniform reaction rate. It is intriguing to improve the utilization degree of the porous electrodes. In this work, a three-dimensional computational model is developed for MFC with flow-through porous electrodes. Characteristics of the reaction rate distributions under different electrode geometries are examined. The results show that reaction rate varies noticeably along the electrode width direction, but minimally along the electrode length direction. High reaction rate region locates in the vicinity of the interface between the porous electrode and the middle channel. A relatively high aspect ratio, defined as the ratio of the electrode length to width, is beneficial to improve the utilization degree of the porous electrodes. Yet, concentration losses increase due to the decreased fluid velocity. Considering the cell performance, optimal electrode aspect ratios are derived for the anode and cathode, respectively.

Keywords: Microfluidic fuel cell; utilization degree; porous electrode.

§Corresponding author.

#To cite this article, please refer to its earlier version published in the *International Journal of Modern Physics B*, Volume 34, 2040022 (2020), DOI: 10.1142/S0217979220400226.

1. Introduction

Nowadays, fuel cells are receiving great attention as a promising green technology to store and supply electricity for mobile applications.[1,2] In the transport sector, major automakers are getting ready to introduce their fuel cell cars to the market.[3,4] Fuel cell is also applicable for portable electronics.[5] However, the technological development is more challenging because better technology is required to lower the cost of portable fuel cell for the highly price-competitive market.[6,7] Microfluidic fuel cell (MFC), also called membraneless laminar flow-based fuel cell, is a promising solution. In MFC, the co-laminar nature of the fuel and oxidant streams in microchannel can segregate the reactants.[8] This unique membraneless feature has advantages including elimination of membrane, high reactant/media flexibility and low cost.[9–11]

To date, a wide array of MFCs has been developed to improve MFC implementations. These MFCs can be divided into two types, i.e., those working with all-liquid reactant solutions and those with a liquid anolyte and oxygen. The possibility of the anolyte/oxygen MFCs for portable electronics is impaired by the yet unsolved problem of low fuel utilization.[12,13] The all-liquid MFCs, therefore, stand out as the most promising candidate due to their much higher fuel utilization when coupled with flow-through porous electrodes. Krishnamurthy *et al.* developed an all-vanadium MFC with flow-through porous electrodes and a fuel utilization as high as 94% was achieved under steady-state operation.[14] Porous electrode is beneficial to the fuel cell systems due to its high surface-to-volume ratio.[11,15–17] To improve the cell performance, it is optimal to operate porous electrode at the maximum reaction rate everywhere within the porous medium. Yet, our previous studies show that porous electrodes in MFCs frequently operate with nonuniform reaction rates.[18] Under certain operation conditions, the reaction rate in some parts of the porous electrodes can be as low as zero.

In this work, a three-dimensional computational model is developed for MFCs with flow-through porous electrodes. Parametric analysis is performed to analyze the geometrical effects in MFCs with flow-through porous electrodes. Characteristics of the reaction rate distributions under different electrode geometries are investigated and optimal electrode aspect ratios are derived for the anode and cathode, respectively.

2. Numerical Model

A three-dimensional numerical model for an all-vanadium MFC with flow-through porous electrodes has been developed in this work to investigate

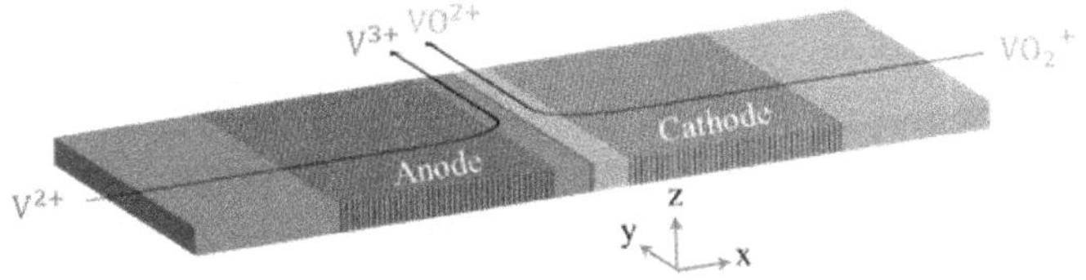

Fig. 1. (Color online) A schematic view of MFC with flow-through porous electrodes.

the geometrical effects in MFCs. The computational domain is shown in Fig. 1. In the operation, the anolyte and catholyte are pumped into the cell from the two opposite sides and then flow face-to-face through the corresponding porous electrodes. When the reactant solutions meet each other in the middle channel, they will flow out of the cell in the laminar flow way.

In the simulation, the mass conservation in the MFC with flow-through porous electrodes is described by the continuity equation. The momentum transport in the regions outside the porous electrodes is determined by the Navier–Stokes equation. The Brinkman equation is used to simulate the momentum transport in the porous electrodes. The species transport of the ions involved in the electrochemical reactions is determined by the species conservation equations. The current density distribution is derived from the charge conservation equations and the charge transfer current density is given by the Bulter–Volmer equation.

A commercial software, COMSOL Multiphysics 5.3a is used to solve the present model at the given operating conditions. Structural grids are generated and grid independence check is performed to assure the accuracy of the numerical solution. The relative convergence tolerance is set to be 1×10^{-4}. Multifrontal Massively Parallel sparse direct Solver (MUMPS) is employed to compute the stationary problem. The final current density is normalized by the anode volume for the convenience of comparison among different cases.

3. Results and Discussion

3.1. *Model validation*

Reliability of the present model is verified by comparing the simulated polarization curve with the experimental data in Ref. 14. The geometrical dimensions and operating conditions of the simulation follow those in the experiment, where 1.89 mol L^{-1} V^{2+} in 4 mol L^{-1} sulfuric acid is used

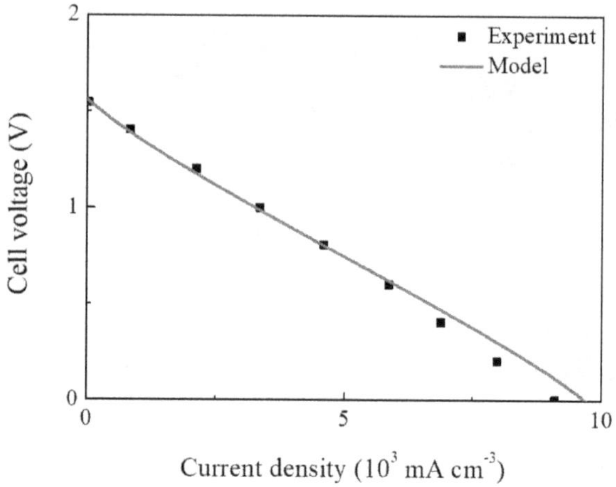

Fig. 2. (Color online) Comparison of experimental and model-derived polarization curves for MFC with flow-through porous electrodes.

as the anolyte, 1.84 mol L^{-1} VO_2^+ in 4 mol L^{-1} sulfuric acid is used as the catholyte, Toray carbon paper (TGPH-090) is employed as the porous electrode material, and flow rate is controlled at 60 μL min^{-1}. Related results are plotted in Fig. 2. It can be seen that the modeling and experimental results fit well with each other, indicating a high reliability of the present model.

3.2. *Reaction rate distribution*

The reaction rate distribution in the present MFC is shown in Fig. 3. The variations of the reaction rate along the electrode length (y-direction) and electrode height (z-direction) are negligible due to the homogeneous electrode properties. Nonuniform reaction rate distribution is observed along the electrode width (x-direction). As shown in the anode, the reaction rate decreases obviously along the negative x-direction and the peak reaction rate occurs at the anode/channel interface, which is mainly attributed to the increased ohmic losses. In the cathode, similar reaction rate distribution can be observed, where the reaction rate decreases along the positive x-direction and the peak reaction rate occurs at the cathode/channel interface.

Considering the low reaction rate in the vicinity of the electrode inlets, a relatively high aspect ratio, defined as the ratio of the electrode length to

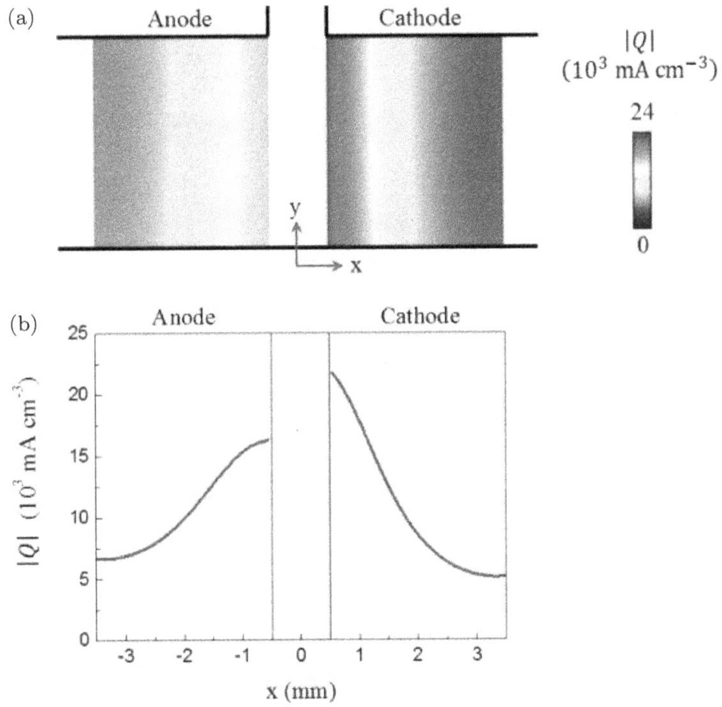

Fig. 3. (Color online) Reaction rate distributions (a) in the mid-plane and (b) along the mid-line of the porous electrodes.

width, is beneficial to improve the utilization degree of the porous electrode. Yet, if we increase the electrode aspect ratio by increasing the electrode length, concentration losses will increase due to the decreased fluid velocity. Therefore, decreasing the electrode width is an optimal way to increase the electrode aspect ratio. Besides, it is worth mentioning that the variation range of the reaction rate in the cathode is larger compared to that in the anode as the VO_2^+ reduction reaction has faster electrochemical kinetics than the V^{2+} oxidation reaction. Thus, the optimal electrode aspect ratio should be different for the anode and cathode due to the anode–cathode mismatch on the electrochemical kinetics.

3.3. *Electrode geometry studies*

From Sec. 3.2, we know that the variation of the reaction rate distribution is mainly controlled by the electrode width in the x-direction. Thus in the

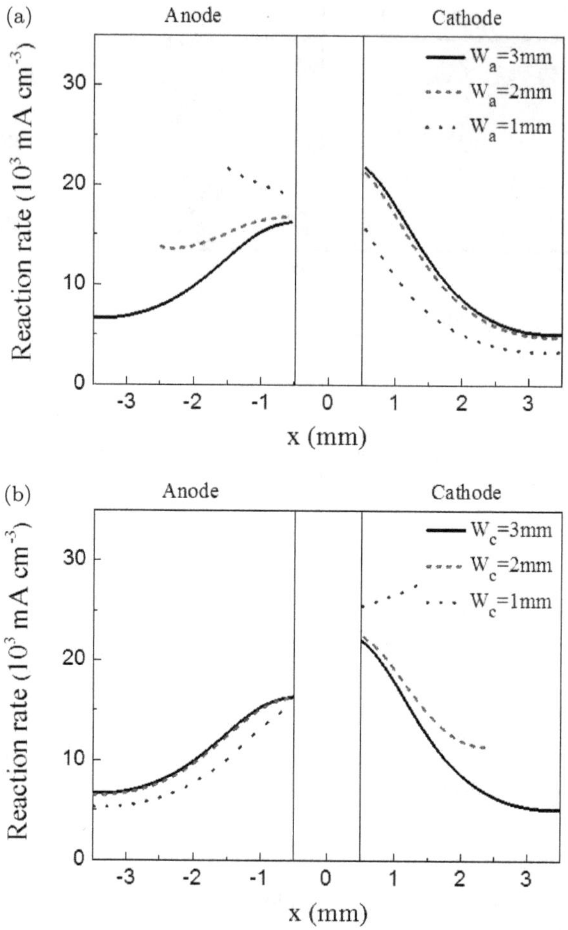

Fig. 4. (Color online) Effects of (a) anode width and (b) cathode width on the reaction rate distribution.

following electrode geometry studies, the electrode aspect ratio is shifted by decreasing the electrode width. The effect of the anode aspect ratio is examined first. As shown in Fig. 4(a), when the anode width is decreased from 3 to 2 mm, the reaction rate distribution in the cathode remains almost the same, indicating that the VO_2^+ reduction reaction in the cathode is not influenced by the change in the anode width. However, when we further decrease the anode width from 2 to 1 mm, obvious decrease is observed in the cathode reaction rate. Thus, optimal value exists for the anode aspect

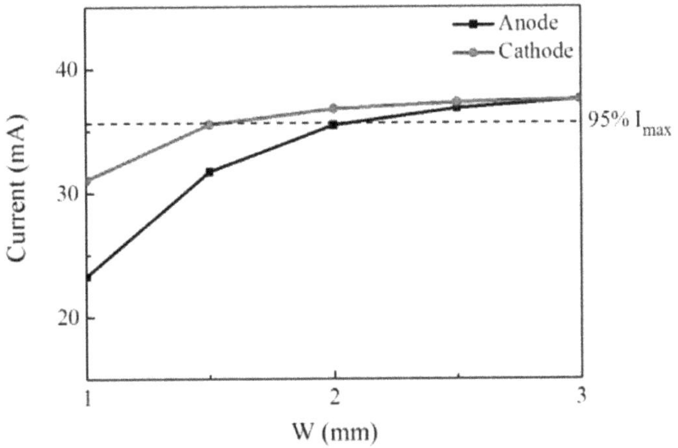

Fig. 5. (Color online) Effects of the electrode width on the output current.

ratio. In Fig. 4(b), the effect of the cathode aspect ratio is also examined and similar phenomenon is observed. As the VO_2^+ reduction reaction has faster electrochemical kinetics than the V^{2+} oxidation reaction, the effect of the cathode aspect ratio is not as significant as that of the anode.

In Fig. 5, the effect of the electrode aspect ratio on the cell performance is presented. From the results, the output current first increases rapidly with the increasing electrode width, but then the increasing trend becomes weak. When the electrode width increases to a certain value, further increasing the electrode width does not yield any obvious change in the output current. For the anode, when the anode width increases to 2 mm, the output current is already 95% of the maximum value obtained at exceedingly large anode width. Thus, optimal electrode width for the anode is 2 mm and corresponding anode aspect ratio is 2. Similarly, the optimal electrode aspect ratio for the cathode is obtained, which is 4 due to its faster electrochemical kinetics.

4. Conclusions

In this work, a three-dimensional computational model is developed to investigate the geometrical effects in MFC with flow-through porous electrodes. Characteristics of the reaction rate distributions under different electrode geometries are examined. The results show that porous electrodes in MFC operate with nonuniform reaction rate and optimal values exist for the electrode aspect ratio. Considering that the electrochemical kinetics of the V^{2+} oxidation reaction are slower than those of the VO_2^+ reduction reaction, different optimal electrode aspect ratios are derived for the anode and cathode, which are 2 and 4, respectively.

Acknowledgments

The research is supported by the Natural Science Foundation of Jiangsu Province (BK20170314), Changzhou Science & Technology Program (CJ20179016) and Changzhou Institute of Technology 2016 annual scientific research fund project (YN1604).

References

1. A. Amiri *et al.*, *Int. J. Hydrogen Energy* **43**, 23475 (2018).
2. Y. Wang *et al.*, *Appl. Energy* **203**, 535 (2017).
3. S. Hardman and G. Tal, *Int. J. Hydrogen Energy* **43**, 17857 (2018).
4. K. Song *et al.*, *J. Power Sour.* **389**, 230 (2018).
5. L. Li *et al.*, *Appl. Energy* **206**, 413 (2017).
6. Y. H. Kwok *et al.*, *J. Power Sour.* **409**, 58 (2019).
7. M. Tanveer and K.-Y. Kim, *Energy Convers. Manage* **166**, 328 (2018).
8. L. Li *et al.*, *Electrochim. Acta* **187**, 636 (2016).
9. Z. Liu *et al.*, *Int. J. Hydrogen Energy* **43**, 22467 (2018).
10. C. A. Martins *et al.*, *Electrochim. Acta* **271**, 537 (2018).
11. Y. H. Kwok *et al.*, *Appl. Energy* **217**, 258 (2018).
12. H. Xu *et al.*, *Appl. Energy* **160**, 930 (2015).
13. A. K. Rathoure and H. Pramanik, *Int. J. Hydrogen Energy* **41**, 15287 (2016).
14. D. Krishnamurthy *et al.*, *J. Power Sour.* **196**, 10019 (2011).
15. Y. H. Kwok *et al.*, *J Power Sour.* **349**, 75 (2017).
16. H. Zhang *et al.*, *J. Power Sour.* **363**, 61 (2017).
17. Y. H. Kwok *et al.*, *Energy Proc.* **142**, 1522 (2017).
18. L. Li *et al.*, *Energy* **88**, 563 (2015).

https://doi.org/10.1142/9789819818433_0013

Chapter 13

Surface Modification of TiO_2 for Visible Light Photocatalysis: Experimental and Theoretical Calculations of Its Electronic and Optical Properties[#]

Dessy Ariyanti[*,†,‖], Surayya Mukhtar[‡,**], Nisar Ahmed[§,††], Zhuofeng Liu[¶,‡‡], Junzhe Dong[†,§§] and Wei Gao[†,¶¶]

[*] *Department of Chemical Engineering, Universitas Diponegoro, Semarang 50275, Indonesia*

[†] *Department of Chemical and Material Engineering, The University of Auckland, Auckland 1142, New Zealand*

[‡] *Department of Physics, Allama Iqbal Open University, Islamabad, Pakistan*

[§] *Department of Physics and Applied Mathematics, Pakistan Institute of Engineering and Applied Sciences, Nilore, Islamabad, Pakistan*

[¶] *College of Aerospace and Materials Engineering, National University of Defence Technology, Changsha 410073, P. R. China*

[‖] *dessy.ariyanti@che.undip.ac.id*
[**] *surayyaafridi@yahoo.com*
[††] *nisarbinwali@gmail.com*
[‡‡] *liuzhuofeng@hotmail.com*
[§§] *jdon296@aucklanduni.ac.nz*
[¶¶] *w.gao@auckland.ac.nz*

Surface modification has been used as a method to create defects on TiO_2 materials, which can improve their desirable properties. In this paper, defected TiO_2 nano-powder was successfully synthesized by chemical reduction using $NaBH_4$ as the reducing agent at 300–400°C under argon atmosphere.

[‖] Corresponding author.

[#] To cite this article, please refer to its earlier version published in the *International Journal of Modern Physics B*, Volume 34, 2040067 (2020), DOI: 10.1142/S0217979220400676.

High defect concentration can be produced by increasing process temperature. The modified TiO_2 shows good visible light absorption and photocatalytic activity on degradation of Rhodamine B (4–9 times higher than the pristine TiO_2) with the visible light irradiation. Further XPS analysis and theoretical studies using full potential linearized augmented plane wave (FP-LAPW) method as implemented in wien2k code revealed the existence of oxygen vacancy and Ti^{3+} in the modified samples. These types of defects were responsible for the modifications of the electronic and optical properties of TiO_2, resulting in the improved photocatalytic activity in visible light irradiation.

Keywords: $NaBH_4$ reduction; defect; density functional theory; TiO_2; visible light absorption.

1. Introduction

TiO_2 is a wide bandgap semiconductor (3.0–3.2 eV) that can be used as photocatalyst.[1] It requires ultraviolet radiation (290–400 nm) to excite electrons from the valance band, not visible or fluorescent light,[2] which limit their application in utilizing major solar light spectrum. In order to enable light absorption in the visible region and to improve the photo-efficiency of TiO_2, doping TiO_2 with nonmetals (such as C, N, S or F), metals (such as Fe, Cr or V) or metal oxides (such as V_2O_5, ZrO_2 or WO_3) has been reported. Hybrid materials combining TiO_2 with carbon materials such as carbon nanotubes, activated carbons or TiO_2-carbon microspheres were also investigated for the same propose.[3–5] In addition, new photocatalytic materials have been developed to achieve higher activities than TiO_2-based photocatalysts; but the success cases for practical applications are few. One of the most viable and practical approaches in developing better photocatalysts is to use self-modify TiO_2 processes, including surface modification.

In this paper, surface modification has been conducted by chemical reduction using $NaBH_4$ to create defects in pristine TiO_2 powder. $NaBH_4$ is an inorganic compound that can be used as a versatile reducing agent. Thermal decomposition of $NaBH_4$ releases active hydrogen that is favorable for defect introduction to the surface of semiconductor.[6] The effect of temperature was studied in correlation to its optical and electronic properties. The full potential linearized augmented plane wave (FP-LAPW) method within density functional theory (DFT) was employed to investigate the effect of oxygen vacancy on the optical and electronic properties of the modified samples.

2. Experimental Details

2.1. *Preparation of modified TiO_2 and sample characterization*

Anatase TiO_2 (Sigma Aldrich Co. size <25 nm) and reduction agent $NaBH_4$ (ECP Ltd.) were used as the starting material. First, 0.75 g $NaBH_4$ was mixed with 2.0 g TiO_2, heating then applied for 60 min in a furnace under Ar to 300–400°C (labelled X300, X350 and X400, cooled down, washed with DI water, and dried at 110°C.[7] The prepared samples were then characterized by XRD, UV-Vis-NIR and XPS.

2.2. *Photocatalytic degradation of organic pollutant*

Photocatalytic activity was evaluated by monitoring the decomposition of RhB in an aqueous solution under visible light irradiation (ULTRA VITALUX 300 W 230 V E27 with UV cut-off filter). Photocatalyst (50 mg) was mixed with RhB solution (50 mL, 5 ppm), followed by 30 min stirring in dark to reach the adsorption equilibrium. Two milliliters were taken every time to filter by 0.2 mm PVDF syringe filter. The concentration of RhB was determined with a UV-Vis spectrophotometer by the peak intensity at 553 nm.

2.3. *Method for theoretical calculations*

The calculations were first carried out on the primitive unit cell of TiO_2 in anatase phase. The $2 \times 2 \times 2$ 96-atom anatase supercell was used to construct TiO_2 systems with oxygen vacancy V_o (concentration was set at 4.1%) and to calculate the V_o energy formation using FP-LAPW method as implemented in wien2k code.[8] The wavefunction for core states is taken as atomic like function, while for the valence states, it is considered as plane waves. The muffin-tin radius for core states is denoted as R_{MT}. Outside R_{MT}, valence states are accounted. In this work, nonoverlapping R_{MT} for Ti and O atoms were 1.93 and 1.75 a.u, respectively. Energy convergence criteria were set at 10^{-4} eV. For calculation of electronic and optical properties, a k mesh of 36 points was sampled in the irreducible Brillouin zone (IBZ). The maximum cut-off for the plane wave k_{max} is 6.5. Exchange correlations effects are considered under generalized gradient approximation (GGA), GGA+U (with U as Hubbard potential, set as 10.303 eV) and Tran–Blaha modified Becke–Johnson (TB-mBJ) approximation.

3. Results and Discussion

3.1. *Structures features and physical properties*

As shown in Fig. 1(a), small changes such as peak boardening and reduction of the peak intensity were observed in the XRD peaks of the modified TiO_2 in comparison to the pristine one. The diffraction peaks of all samples were assigned for TiO_2 anatase (JCPDS card 78-2486) indicating that all samples have the same crystalline structure (anatase). Only peak intensity was reduced gradually with increasing. The deformation of cystalline lattice may occur and defect be created, thus resulting in low peak intensity.[7] Table 1 shows the crystallite size and lattice strain of the pristine and $NaBH_4$ modified TiO_2 estimated by Debye–Scherrer and Stokes–Wilson equations. As the temperature of reduction process gets higher, the crystallite size slightly reduces (from 63.5 to 48.8 nm), and the lattice strain slightly increases (0.010–0.012). Strain broadening is often triggered by lattice imperfection, including vacancies, interstitials and substitutions.[9]

Figure 1(b) depicts the optical properties of pristine and modified TiO_2 samples which are represented by the light absorption behavior. The absorbance intensity from 400 to 800 nm increased as the $NaBH_4$ treatment temperature increased. All modified samples were able to absorb visible light while pristine TiO_2 has no absorption. The enhancement of light visible light absorption in TiO_2 were also reported in Refs. 6 and 10–13.

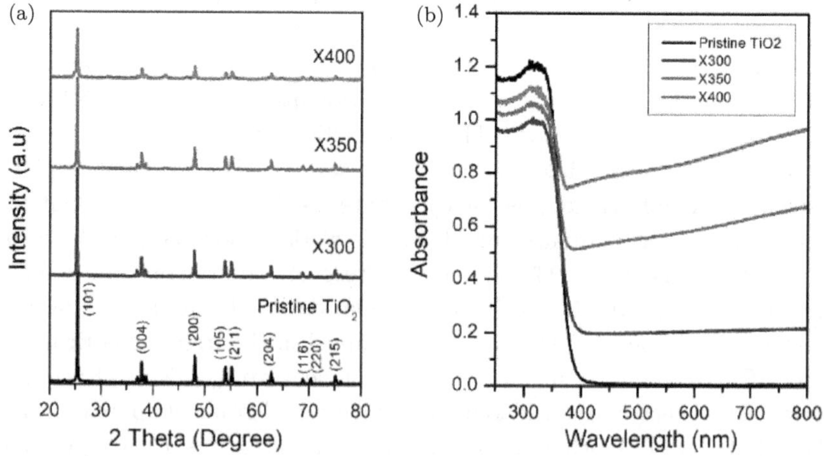

Fig. 1. (Color online) (a) XRD pattern and (b) UV-Vis spectrum of pristine and $NaBH_4$ modified TiO_2.

Table 1. Crystallite size and lattice strain of pristine and $NaBH_4$ modified TiO_2.

Samples	FWHM	Peak position	Crystallite size (nm)	Lattice strain
TiO_2 pristine	0.128	25.41	63.5	0.00991
X300	0.126	25.22	61.6	0.00981
X350	0.134	25.26	60.2	0.01044
X400	0.157	25.24	48.8	0.01224

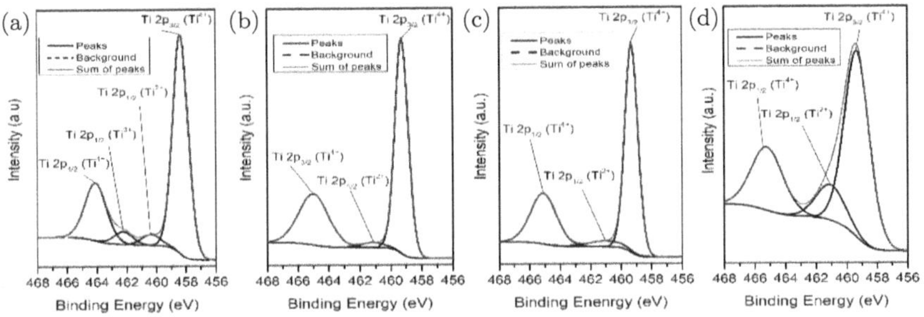

Fig. 2. (Color online) XPS spectra of Ti 2*p*: (a) pristine TiO_2, (b) X300, (c) X350 and (d) X400.

Table 2. Percentage area of O 1*s*.

Sample	O 1*s* (Ti-O) (%)	O 1*s* (O-H) (%)
Pristine TiO_2	68.10	31.90
X300	62.65	37.35
X350	56.26	43.74
X400	44.54	56.46

Figure 2 compares the XPS spectra of Ti 2*p* of pristine and modified TiO_2 at different temperatures. Ti^{4+} and Ti^{2+} were detected in all samples, representing the same state of Ti atoms that existed in these samples. However, additional peak of Ti^{3+} was detected in X350 [Fig. 2(c)], indicating the existence of defects. Hydroxide (O-H) and Ti (Ti-O) in the oxygen state (O 1*s*) were also detected. The presence of hydroxide is predictable as the defect formation involves the attraction of active hydrogen with the oxygen in the lattice structures, thus it can be associated with the surface defects and oxygen vacancies.[14] Table 2 shows the percentage of O 1*s* (H-O) in the sample, which increases gradually with increasing temperature. High percentage of O 1*s* (H-O) indicates the existence of distortion on the TiO_2 surface and more severe conditions compared to the distortion in bulk.[7]

XPS also indicate the electronic band position of the materials (valence band). The valence band maximum (VBM) is estimated by a linear extrapolation of the peak to the baseline.[15] The pristine TiO_2 have a band edge position of 1.7 eV below the Fermi energy. Meanwhile, the band edge of modified TiO_2 were slightly lower by 0.1; 0.4 and 0.2 eV for X300; X350 and X400, respectively. There is a possibility of the formation of an extra band between the Fermi level (0 eV) and the top of the VB triggered by the defect of Ti^{3+} and oxygen vacancies.[15] To confirm the formation of extra band of the modified TiO_2, the theoretical calculation based on FP-LAPW DFT was conducted and discussed in Sec. 3.3.

3.2. *Photocatalytic activity*

Table 3 presents the kinetic constant of pristine and modified TiO_2 on RhB degradation under visible light. Sample X350 has the highest degradation rate of $k = 0.0018$/min, followed by X300 and X400 with $k = 0.0011$ and 0.0008/min, respectively. The pristine TiO_2 has the lowest degradation rate as it does not absorb visible light. As X400 has higher oxygen vacancies O 1s (H-O), it has lower photocatalytic activity compared to X300 and X350. This can be explained by the possibility of samples having different type and concentration of defects. By decreasing the relative defect concentration of bulk to surface in the TiO_2, the photocatalytic efficiency is significantly enhanced.[14,16]

3.3. *Theoretical studies*

The calculated bandgap energy for pristine TiO_2 by using GGA, GGA+U and TB-mBJ approximations is 2.015, 3.246 and 2.872 eV, respectively. Figure 3(a) shows the total density of states for anatase TiO_2 unit cell. Based on Fig. 3(a), the conduction band is made up of Ti-3d valence states while O-2p states contribute in the formation of the valence band.

Table 3. Kinetics constant of RhB degradation under visible light.

Catalyst	Kinetics constant (min^{-1})	R^2
Pristine TiO_2	0.0002	0.9601
X300	0.0011	0.9648
X350	0.0018	0.9811
X400	0.0008	0.9864

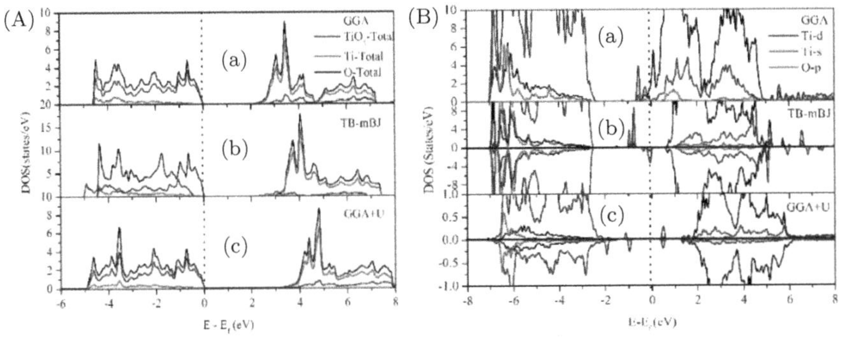

Fig. 3. (Color online) Density of states of (a) pristine and (b) defected TiO_2 with oxygen vacancy V_O.

The supercell model, oxygen vacancy V_o, was created; and the effect of V_o on the bandgap and state of density of states was investigated. Density of states (DOS) of Ti-d, Ti-s and O-p states are plotted [Fig. 3(b)]. It shows the new localized states between the conduction band and valence band are O-2p states which are created due to the V_o defect. TB-mBJ gives bandgap ~0.7 eV in spin up channel and ~0.446 eV in spin down channels. The conduction band and valence band both shift to the lower energy levels and become narrower. DOS shows that new O-2p states are located at 1.1 eV above the VBM due to V_o defect. Similar to GGA, these localized energy states in the valence band act as the trap centers for photoexcited holes.

GGA+U approximation depicts a different picture as compared to GGA and TB-mBJ schemes. The bandgap is improved and gives ~1.5 eV in spin down channel whereas it reduced to ~0.426 eV in the spin up channel. The energy states due to V_o are located in the conduction band and valence band. These are O-2p states which form shallow levels at the CBM and localized energy states at 1.5 eV below the conduction band. In the valence band, they form shallow donor energy levels at VBM edge and localized energy states at ~0.7 eV above the VBM. The possible explanation of this behavior is that V_o acts like donor impurity with a charge state +2. With the creation of V_o, three O-2p states but four electrons were removed, which are responsible for the valence band formation. Now, two Ti-3d electrons need to transfer from the conduction band to valence band to fill the O-2p_y and O-2p_z states in the pristine TiO_2 unit cell, residing in the conduction band.

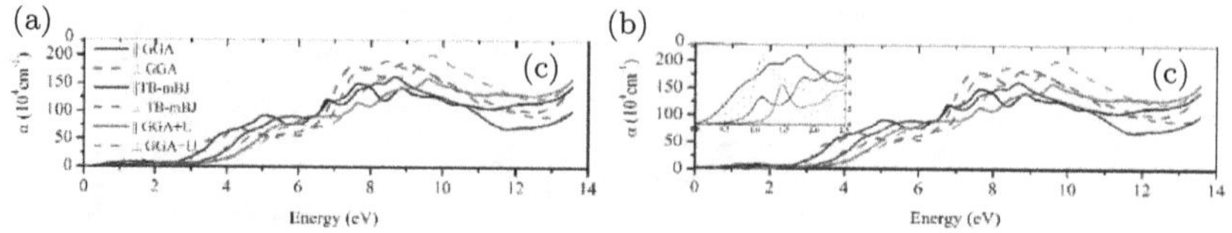

Fig. 4. (Color online) Absorption coefficient spectra of (a) pristine and (b) defected TiO_2.

These unpaired Ti-3d electrons behave like free electrons in the conduction band, and are responsible for the n-type conductivity in TiO_2. Reports cited the effect of V_o defects on the photocatalytic activity.[17] The presence of oxygen vacancies in TiO_2 can increase or reduce the photocatalytic activity depending on the factors of the concentration of V_o, oxygen vacancy at the surface or in the bulk. They provide trap centers for electrons and holes, either increase or decrease the electron–hole recombination rate. It is also reported that the electrons can easily be photo-excited to these localized states by visible light absorption and cause an increase in the photocatalytic activity.[17]

The threshold energy for absorption is closely related to the optical bandgap of the material. For pristine TiO_2, it is calculated as 2.3–2.5, 3.2–3.5 and 3.5–4.0 eV in parallel and perpendicular directions of incident photons with GGA, TB-mBJ and GGA+U approximations, respectively [Fig. 4(a)]. Figure 4(b) shows the light absorption in the visible energy range of ~0.5–2.5 eV for defected TiO_2 and confirmed by the experimental results presented in Fig. 2. Although the absorption in visible light is very low compared to the UV region, the photocatalytic activity is significantly enhanced by V_o defects, as shown in Table 3.

4. Conclusion

Defects have been successfully introduced to the pristine TiO_2 via $NaBH_4$ treatment. Its concentration can be controlled by temperature. The modified TiO_2 has good visible light absorption and high photocatalytic activity towards RhB degradation (4–9 times higher than pristine TiO_2). This may come from the formation of V_o and Ti^{3+}. Theoretical calculations confirm the presence of extra energy states created by V_o. The bandgap is reduced in the TiO_2 with V_o defects as compared to pristine TiO_2. These changes lead to a disordered layer and a broad visible light absorption which is in line with the experimental findings.

References

1. A. Fujishima and K. Honda, *Nature* **238**, 37 (1972).
2. X. Wang *et al.*, *J. Hazard. Mater.* **262**, 16 (2013).
3. M. Sillanpää, M. C. Ncibi and A. Matilainen, *J. Environ. Manage* **208**, 56 (2018).
4. A. Fujishima, X. Zhang and D. A. Tryk, *Surf. Sci. Rep.* **63**, 515 (2008).
5. X. Chen, L. Liu and F. Huang. *Chem. Soc. Rev.* **44**, 1861 (2015).
6. H. Tan *et al.*, *Nanoscale* **6**, 10216 (2014).
7. D. Ariyanti *et al.*, *Mater. Chem. Phys.* **199**, 571 (2017).
8. P. K. S. Blaha *et al.*, WIEN2k: An augmented plane wave plus local orbitals program for calculating crystal properties (2001).
9. D. Balzar, *J. Res. Natl. Inst. Stand. Technol.* **98**, 321 (1993).
10. J. Dong *et al.*, *ACS Appl. Mater. Interfaces* **6**, 1385 (ẏ2014).
11. H. Wang *et al.*, *Catal. Commun.* **60**, 55 (2015).
12. G. Zhu *et al.*, *ChemCatChem.* **7**, 2614 (2015).
13. T. Lin *et al.*, *Energy Environ. Sci.* **7**, 967 (2014).
14. J. Yan *et al.*, *Phys. Chem. Chem. Phys.* **15**, 10978 (2013).
15. S. Pan *et al.*, *J. Mater. Chem. A* **3**, 11437 (2015).
16. M. Kong *et al.*, *J. Am. Chem. Soc.* **133**, 16414 (2011).
17. X. Pan *et al.*, *Nanoscale* **5**, 3601 (2013).

https://doi.org/10.1142/9789819818433_0014

Chapter 14

Study of Fast Charging Battery[#]

Mingfang Liu*, Feng Rao, Bei Li, Xinya Jiang† and Xianggang Meng

Changzhou Institute of Technology,
Liaohe Road 666, Changzhou, Jiangsu, 213000, P. R. China
**summerlmf@126.com*
†58875830@qq.com

The key to battery management systems (BMS) is an accurate and real-time prediction on State of Charge (SOC) of the power battery. The methods of estimating SOC of power battery were analyzed. The gray neural network model was introduced into the field of SOC estimation. The model is used to establish the relationship of SOC, discharge current and rebound voltage. The model can be used to estimate SOC of battery on line by detecting the values of rebound voltage and discharge current during the discharge process. The model of SOC estimation is proved feasible by comparing the experimental data with calculated data.

Keywords: Power battery; rebound voltage; SOC estimation.

1. Introduction

In recent years, electric vehicle (EV) has attracted worldwide attention because of its advantages of energy efficiency and environmental friendliness. As one of the key technologies of EV, battery management technology has become a key research topic all over the world. The battery management system accounts for more than 1/3 of the whole battery system in cost. In effect, the improvement of battery management technology can improve the overall performance of battery system, so it has become a new research and development field of EV.[1,2]

*Corresponding author.

[#]To cite this article, please refer to its earlier version published in the *International Journal of Modern Physics B*, Volume 34, 2040023 (2020), DOI: 10.1142/S0217979220400238.

Power battery State of Charge (SOC) is an important parameter of power battery status. It plays an important role in real-time accurate battery estimation, state monitoring, battery life improvement and power supply security. Accurate estimation of battery SOC can grasp the status of batteries in real-time, prevent overcharge and over-discharge during the operation of batteries, help to protect batteries and prolong the service life of batteries.[3–5]

2. Gray Neural Network Model

In this paper, the gray model and the neural network are combined, and the embedded model is selected to construct the prediction model. The modeling steps are as follows:[1–5]

(1) First, GM $(1,3)$ model is used to predict and get the data reduction value $\widehat{x}^{(0)}(k)$. Set $e^{(0)}(k)$ as a residual sequence, and $e^{(0)}(k) = x^{(0)}(k) - \widehat{x}^{(0)}(k)$.
(2) BP network model of residual sequence is established. If the prediction order is s, the $e^{(0)}(k-1), e^{(0)}(k-2), \ldots, e^{(0)}(k-s)$ is used to predict the time value. The input sample of BP network training is $e^{(0)}(k-1)$, $e^{(0)}(k-2), \ldots, e^{(0)}(k-s)$, and the $e^{(0)}(k)$ value is the expected value of prediction. The trained BP neural network can be used to predict the residual sequence.
(3) A new prediction sequence $\{\widehat{y}^{(0)}(k)\}$ is constructed. New predicted values $\{\widehat{e}^{(0)}(k)\}$ are obtained from step (2) and $\widehat{y}^{(0)}(k) = \widehat{x}^{(0)}(k) + \widehat{e}^{(0)}(k)$.

3. Theoretical Analysis of Rebound Voltage

The voltage at both ends of the battery will jump at the moment of disconnection, as shown in point A to point B in Fig. 1. The point at which the battery discharges to the cut-off voltage is recorded as point A, and the discharge circuit is cut off. The terminal voltage of the battery will jump from point A to point B. After it is offline, point C is the voltage of the battery, and will be restored to a relatively stable value after a period of stationary. The voltage of point B is called the rebound voltage, which is denoted as U_t.

The mathematical model of the battery is shown in Fig. 1. The dynamic process from point B to point C is analyzed according to the mathematical

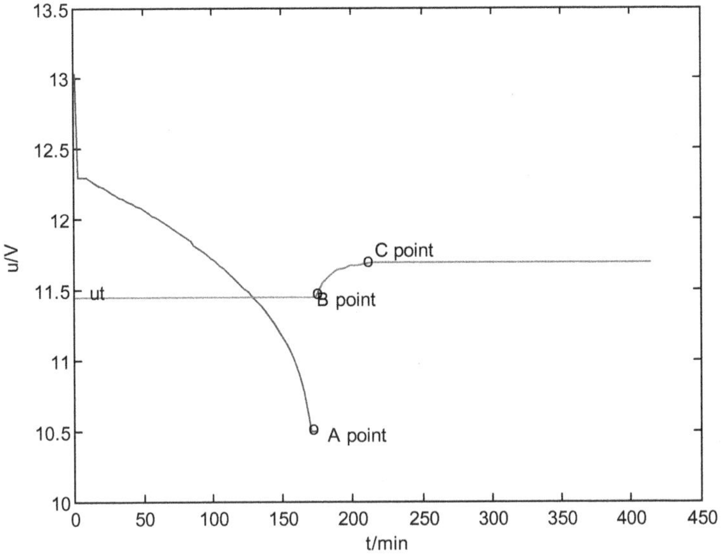

Fig. 1. (Color online) Rebound voltage of power battery after discharge.

model diagram

$$U_{\mathrm{oc}}(\mathrm{s}) - U_{\mathrm{cell}}(\mathrm{s}) = I(s) \times \left(R_s + \frac{R_{ts} \times \frac{1}{C_{ts}}}{R_{ts} + \frac{1}{C_{ts}}} + \frac{R_{tl} \times \frac{1}{C_{tl}}}{R_{tl} + \frac{1}{C_{tl}}} \right),$$

where R_s is the resistance of the loop, R_{ts} and C_{ts} are the time constant of short-term transient response and R_{tl}, C_{tl} are the time constant long-term transient response.

After pull-type inverse transformation, the result is

$$\begin{aligned} U_{\mathrm{oc}}(t) - U_{\mathrm{cell}}(t) &= i(t) \times R_s + i(t) \times R_{ts} \times (1 - e^{-\frac{t}{\tau_{ts}}}) \\ &\quad + i(t) \times R_{tl} \times (1 - e^{-\frac{t}{\tau_{tl}}}) \end{aligned}$$

where $\tau_{ts} = R_{ts} \times C_{ts}$, $\tau_{tl} = R_{tl} \times C_{tl}$; then, the voltage of both batteries is

$$\begin{aligned} U_{\mathrm{cell}}(t) &= U_{\mathrm{oc}}(t) - i(t) \times (R_s + R_{ts} + R_{tl}) \\ &\quad + i(t) \times (R_{ts} \times e^{-\frac{t}{\tau_{ts}}} + R_{tl} \times e^{-\frac{t}{\tau_{tl}}}). \end{aligned}$$

It is simplified

$$U_{\mathrm{cell}}(t) = U_{\mathrm{oc}}(t) - U_1(t) + U_2(t),$$

where $U_1(t)$ is produced by the resistance in the circuit.

When the discharge stops, the current can be expressed as

$$i(t) = \begin{cases} K, t < 0 \\ 0, t \geq 0 \end{cases} \quad t = 0,$$

$U_1(t) = 0$ is the line AB in Fig. 1.

In form 4.4, $U_2(t)$ is the recovery voltage

$$U_2(t) = i(t) \times (R_{ts} \times e^{-\frac{t}{\tau_{ts}}} + R_{tl} \times e^{-\frac{t}{\tau_{tl}}}).$$

It is known when $t > 0$, the exponential decay occurs because the current has a negative jump, which corresponds to the negative input signal. The voltage at the battery terminal presents a slow regression trend as the BC line in Fig. 1. When $t = 0$, the voltage of both batteries is the rebound voltage U_t; when $t \to \infty$. The open-circuit voltage of the battery is equal to the electromotive force of the battery:

$$U_t = U_{\text{cell}}(t)|_{t=0} = U_{\text{oc}}(t) + i(t) \times (R_{ts} \times e^{-\frac{t}{\tau_{ts}}} + R_{tl} \times e^{-\frac{t}{\tau_{tl}}}).$$

When the current is zero, the voltage loss caused by polarization has not been fully restored. After a certain period of time, the voltage of both batteries may be restored to the open-circuit voltage.

From the above analysis, the rebound voltage depends on the battery open-circuit voltage, and the circuit voltage has a known relationship with the SOC of the battery. The relationship between rebound voltage and the SOC can be obtained, and the rebound voltage can be obtained at the instant of being offline.

Based on the relationship between rebound voltage and SOC studied above, the gray model GM $(1, 3)$ is used to predict the SOC of power batteries, and then the residual error of the gray prediction model is corrected by using BP neural network, resulting in a more accurate prediction.

4. Test and Result Analysis

BX7B-3A lead-acid power battery and BTS-M 50A/12V power battery integrated parameters automatic tester were used in this work. Two-stage constant current–constant voltage charging is used in charging experiment, while constant current discharging is used in discharging experiment.

The following experiments were carried out on the battery: at different discharging rates, discharging to different capacity, recording the rebound voltage when the discharging circuit was disconnected; after 2 h, discharging at 0.2 C to 10.5 V, and recording the subsequent discharging capacity as the remaining capacity of the battery.

The data obtained at constant temperature are shown in Table 1. Rebound voltage and discharge rate were selected, and 40 groups of training samples and 6 groups of prediction samples were selected. The data are shown in Table 1.

First, Gray GM (1,3) prediction is carried out, and the response equation of GM (1,3) model is obtained as follows:

$$\begin{aligned}\widehat{x}_1^{(1)}(k+1) &= (75.24 - 0.07x_2^{(1)}(k+1) + 1.19x_3^{(1)}(k+1))e^{-0.69k}\\ &\quad + 0.07x_2^{(1)}(k+1) - 1.19x_3^{(1)}(k+1) \quad k = 0, 1, \ldots, n-1.\end{aligned}$$

Then BP neural network is established by residual sequence. In the calculation of BP network, the number of neurons in input layer is 3, the number of neurons in hidden layer is 5, the number of neurons in output layer is 1, and the learning rate is 0.1. The prediction results are shown in Table 2.

Table 1. Experimental data.

	0.2 C discharge				0.5 C discharge			
Discharge capacity DOD	60%	50%	40%	30%	65.91%	50%	40%	30%
Rebound voltage	11.753	11.926	12.09	12.218	11.234	11.86	12.005	12.20
SOC	19.671	32.786	42. 933	56.343	12.914	30.843	35.243	47.99

	0.8 C discharge			1 C discharge		
Discharge capacity DOD	50%	40%	30%	30%	20%	10%
Rebound voltage	11.609	11.916	12.123	12.343	12.289	12.476
SOC	26.029	38.614	43.514	43.243	53.014	63.457

Table 2. Prediction results of two models.

Prediction sample	Original value (%)	GM (1, 3) model: Predicted value (%)	GM (1, 3) model: Error (%)	Gray neural network model: Predicted value (%)	Gray neural network model: Error (%)
41	74.77	75.65	−0.88	73.52	1.25
42	54.31	62.22	7.91	58.18	−3.87
43	50.96	49.34	1.62	50.06	0.9
44	33.11	22.93	10.18	30.71	2.4
45	28.89	35.62	−6.73	31.98	−3.09
46	15.53	12.52	3.01	14.30	1.23

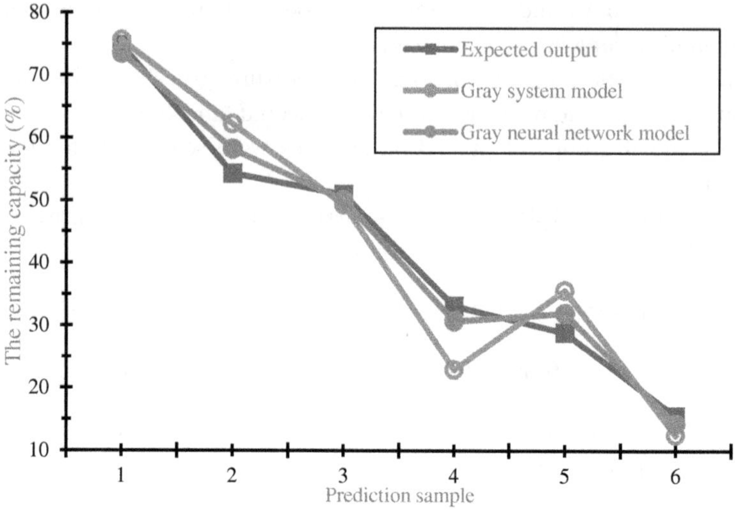

Fig. 2. (Color online) Prediction of SOC gray neural network for batteries.

Table 2 shows that the relative error of the calculated gray neural network is smaller than that of GM (1,3) model in Table 1. The variance formula of the prediction error is

$$\mathrm{MSE} = \frac{1}{n}\sum_{i=1}^{n}(y_i - \widehat{y}_i)^2.$$

The variance of GM (1,3) model prediction error is 0.032 and the gray neural network model prediction error is 0.004, showing that the gray neural network predicts the residual capacity more accurately.

The GM (1,3) model and gray neural network prediction are shown in Fig. 2, exhibiting that the gray model can predict the residual capacity of power battery. But the prediction accuracy of gray neural network is significantly higher than that of gray model.

5. Conclusion

The basic concept of power battery and the significance and difficulties of estimation are introduced in detail. The existing estimation methods of SOC for power battery are analyzed. The gray neural network method was used in this paper, and the modeling principle and detailed steps of gray neural network are introduced. GM (1,3) model and gray neural network model are used to predict and calculate SOC of power battery. By comparing the accuracy of test results, it is found that the prediction accuracy of gray neural network model is higher than that of gray theoretical model; and the calculation accuracy of gray neural network model is more accurate than that of gray theoretical model.

Acknowledgment

This research was supported by the National Key Research and Development Program of China (No. 2017YFF0210002).

References

1. C. H. Wang and H. C. Dong, *Automot. Eng.* **37**, 4 (2015).
2. D. Aurbach *et al.*, *Electrochim. Acta* **47**, 1899 (2002).
3. H. Jeong, J. H. Kim and S. Lee, *J. Mater. Chem.* **20**, 9180 (2010).
4. L. H. Gao and W. Sun, *J. Acad. Arm. For. Eng.* **18**, 8 (2004).
5. S. Junbo, *IEEE Trans. Reliab.* **64**, 82 (2015).

https://doi.org/10.1142/9789819818433_0015

Chapter 15

Cu–TiO_2 Nanocomposite Coatings Prepared from Sol-Enhanced Electrodeposition[#]

Zhen He[*,§,‖], Shengping Zhang[*], Li Yin[*],
Muhammed D. Hayat[†] and Peng Cao[†,‡,¶,‖]

[*]*School of Materials Science and Engineering,*
Jiangsu University of Science and Technology,
Zhenjiang 212003, P. R. China

[†]*Department of Chemical and Materials Engineering,*
University of Auckland, Private Bag 92019,
Auckland 1142, New Zealand

[‡]*MacDiarmid Institute for Advanced Materials and Nanotechnology,*
University of Victoria Wellington,
P. O. Box 600, Wellington, New Zealand

[§]*hezhen0826@outlook.com*
[¶]*p.cao@auckland.ac.nz*

A series of TiO_2-modified Cu coatings were developed via a sol-enhanced electrodeposition process. The effect of TiO_2 sol on the properties of fabricated Cu–TiO_2 nanocomposites is studied. Phase composition and microstructures of Cu–TiO_2 deposits were characterized using X-ray diffraction (XRD), atomic force microscopy (AFM) and transmission electron microscopy (TEM). Mechanical properties of the fabricated Cu–TiO_2 coatings were also examined. The proper addition of TiO_2 sol gives rise to refined coating surface with well-distributed TiO_2 nanoparticles, leading to significantly enhanced properties. The microhardness and wear resistance of the Cu–TiO_2 nanocomposite coating (12 mL/L TiO_2 sol) upsurges owing to the strengthening effect of highly dispersed TiO_2 nanoparticles. We also suggest that the excessive addition of TiO_2 sol should be avoided as it will result in an inferior coating quality with a loose structure for the prepared Cu–TiO_2 deposits.

Keywords: Nanocomposite; TiO_2 sol; electrodeposition; copper.

[‖]Corresponding authors.

[#]To cite this article, please refer to its earlier version published in the *International Journal of Modern Physics B*, Volume 34, 2040038 (2020), DOI: 10.1142/S021797922040038X.

1. Introduction

Along with the growing industrial demand for high-performance coating materials, current research is focused on the enhancement of surface properties of applied engineering coatings.[1–5] Owing to its unique combination of low cost and good electrical- and thermal-conductivity, copper is an extensively applied metallic coating material in a broad range of industrial fields such as electronic engineering and thermal conduction engineering.[6,7] Nevertheless, the scale-up application of copper coatings is limited by its poor hardness, low wear resistance and inferior corrosion resistance under harsh service conditions. Recent research has proposed different methods to modify copper coatings, such as thermomechanical treatment, alloying and composite formation. Among these, electrodeposition is a promising technique to control the properties of copper layers by incorporating a suitable second phase material. Some reports found that incorporating SiC particles gives rise to enhanced microhardness in Cu–SiC composite coatings.[8,9] Thiemig and co-workers also confirmed modified surface morphology and microhardness of the electrodeposited Cu–Al_2O_3 composite coatings.[10]

Many studies have attempted to embed TiO_2 nanoparticles into electrodeposited composite coatings, achieving encouraging improvements in coating microhardness, wear resistance and corrosion resistance, etc.[2,11,12] More recently, a novel sol-enhanced electrodeposition technique was developed to manufacture composite coatings involving pre-synthesized TiO_2 sol, which leads to *in-situ* formation of TiO_2 nanoparticles throughout the co-electrodeposition process.[13,14] This sol-enhanced electrodeposition can improve the dispersion of embedded TiO_2 nanoparticles in the deposited composite coatings. For instance, the use of TiO_2 sol in Ni–P and Au–Ni electrodeposition has been proven to result in impressive enhancement in the mechanical performance of nanocomposite coatings.[13,15] However, no study has ever exploited the potential of the sol-enhanced electrodeposition technique in preparing Cu–TiO_2 coatings. Herein, we systematically studied the electrodeposited Cu–TiO_2 nanocomposite coatings with the usage of TiO_2 sol, aiming to establish a new method for making Cu–TiO_2 coatings with superior performances.

2. Experimental

Analytical grade chemicals and deionized (DI) water were used in sample preparation. Stainless-steel substrates were prepared by several pretreatment steps: (1) grinding and polishing; (2) ultrasonic cleaning in acetone and (3) chemical etching by immersing in the concentrated hydrochloric acid solution. TiO_2 sol was prepared through the methods reported in our previous work.[13]

The electrodeposition of Cu–TiO_2 nanocomposite coatings was conducted in a beaker cell at a constant current of 40 mA/cm^2 at room temperature. The stainless-steel substrate (20 mm $\times$ 30 mm) serves as the anode, while a pure copper sheet (40 mm $\times$ 40 mm) was the cathode. The inter-electrode distance was ~40 mm, and the bath agitation was kept at 300 pm. The electrolyte was made by 0.5 M $CuSO_4$+0.1 M H_2SO_4. Before the electrodeposition, different amounts of prepared TiO_2 sol were carefully added into the electrolytic bath under magnetic stirring.

The crystal structure of Cu–TiO_2 coatings was identified by X-ray Diffraction (XRD, D2 Phaser, Bruker, Germany) at a step size of 0.02°. The surface morphology of coatings was determined using laser scanning confocal microscopy (Axioskop2-MAT, Zeiss, Germany) and atomic force microscopy (AFM, Next II, NT-MDT, Russia). Microstructures were identified by a high-resolution transmission electron microscopy (TEM, TECNAI G2 F20, FEI, USA).

The hardness measurement was performed using a Vickers microhardness test machine. Five indentations were randomly conducted on the surface of each sample, and a load of 50 g was applied for 15 s for each indentation. The tribological tests were carried out employing a micro-tribometer (UMT-2, CETR, USA) at room temperature. In each measurement, the wear test was conducted applying a load of 7 N and total sliding distance of 54 m.

3. Results and Discussion

Figure 1 presents the crystal structure of electrodeposited Cu–TiO_2 coatings. All diffraction peaks can be assigned to copper and a significant texture appears at ~40°. There is no TiO_2 diffraction peak identified in Fig. 1, because the content of TiO_2 generated in the co-deposition process is too low. Meanwhile, the peak intensity and width change with the amount of TiO_2 sol addition. An increased TiO_2 sol leads to broader peaks with less intensity until the TiO_2 sol reaches 12 mL/L, beyond which an opposite tendency is observed with increasing TiO_2 sol addition.

Figure 2 presents the surface morphology of the Cu–TiO_2 nanocomposite coatings. No crack can be seen in these deposits, indicating a good electrodeposition quality. From Figs. 2(a) to 2(d), an increased TiO_2 sol addition significantly refines the coating surface, revealing a rather fine and

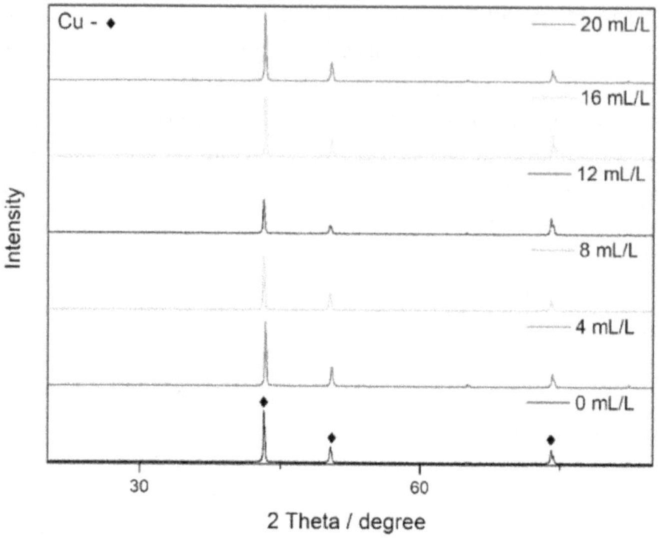

Fig. 1. (Color online) XRD patterns of Cu–TiO_2 nanocomposite coatings prepared at different TiO_2 sol addition.

Fig. 2. Surface morphologies of Cu–TiO_2 coatings prepared at different addition of TiO_2 sol: (a) 0 mL/L, (b) 4 mL/L, (c) 8 mL/L, (d) 12 mL/L, (e) 16 mL/L and (f) 20 mL/L.

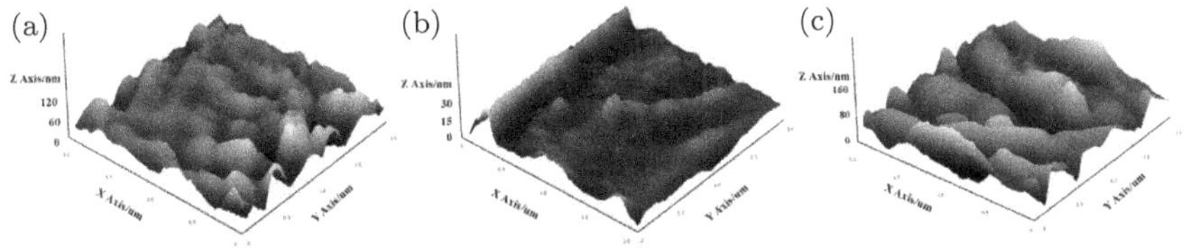

Fig. 3. (Color online) AFM profiles of $Cu–TiO_2$ nanocomposite coatings prepared at different TiO_2 sol addition, (a) 0 mL/L, (b) 12 mL/L and (c) 20 mL/L.

compact surface morphology at 12 mL/L TiO_2 sol addition [Fig. 2(d)]. However, an excessive TiO_2 sol addition at 20 mL/L causes an inferior coating quality with readily identified crystal clusters, Fig. 2(f).

Surface quality of the coating is further revealed by AFM and shown in Fig. 3. When the TiO_2 sol increased to 12 mL/L, the surface became smoother [Fig. 3(b)]. However, further increasing TiO_2 sol caused a rougher and more open surface morphology, Fig. 3(c). In the electrodeposition process of $Cu–TiO_2$ coatings, the well-dispersed TiO_2 nanoparticles facilitate the nucleation process, thus leading to a refined surface morphology with decreased crystal size. On the other hand, the excessive addition of TiO_2 sol may result in aggregation of nanosized particles, which is associated with the inferior coating quality with a rough and open surface.

Neither the structural or morphologic characterization confirm the presence of nanosized TiO_2 due to the detection limit of XRD or SEM, which agrees well with another study.[15] The nanosized particles in the composite coating are difficult to be identified by conventional characterization methods such as XRD and SEM. Here we used a high-resolution TEM to address this challenge and the result is shown in Fig. 4. It can be clearly seen that TiO_2 nanoparticles are distributed in the composite coating with a size of several nanometers. The particle size is much smaller compared with other literature studying nanoparticle-doped composite.[16] This is associated with the continuous *in situ* formation of TiO_2 nanoparticles with a rather fine size, during the electrodeposition process involving presynthesized TiO_2 sol.

The hardness and wear resistance were measured for the prepared $Cu–TiO_2$ nanocomposite coatings, which is summarized in Fig. 5. The coating hardness increases from 106 HV (pure copper layer) to 145 HV upon TiO_2 sol addition of 12 mL/L. Further increasing the TiO_2 sol addition harms the coating hardness. Also, wear resistance measurement reveals a similar trend. The friction coefficient of $Cu–TiO_2$ coatings gradually drops from

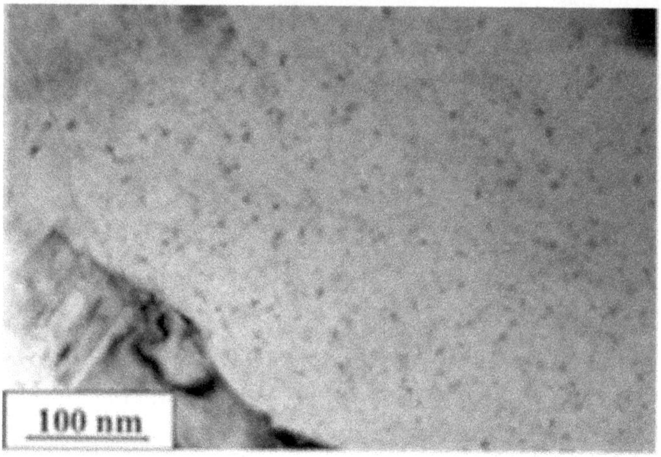

Fig. 4. Microstructure of Cu–TiO_2 nanocomposite observed by TEM.

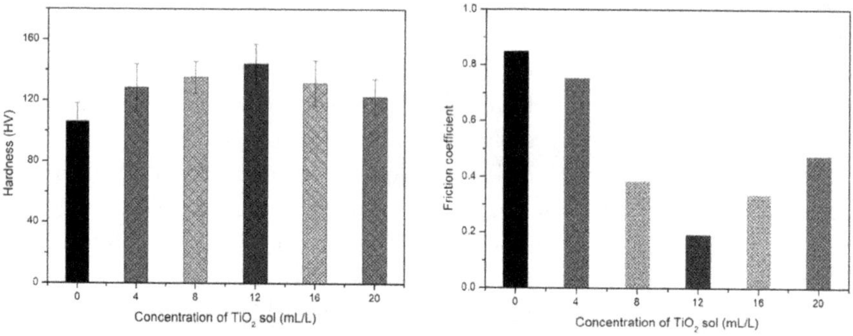

Fig. 5. (Color online) Microhardness and friction coefficient of Cu–TiO_2 nanocomposite coatings prepared at different TiO_2 sol addition.

0.84 of pure copper to 0.19 of Cu-12 mL/L TiO_2 sol, and then increases to 0.46 of Cu-20 mL/L TiO_2 sol. The best mechanical behavior is confirmed for the Cu-12 mL/L TiO_2 sol nanocomposite coating.

At a proper content of TiO_2 sol, the *in situ* generated TiO_2 nanoparticles are highly dispersed in the coating matrix as a second phase. The uniformly distributed TiO_2 nanoparticles in the coating lead to dispersion strengthening effect and grain-refinement strengthening effect, therefore resulting in the improved hardness and wear resistance. However, the increasing agglomeration of TiO_2 particles in the coating causes decreased strengthening effect and a loose structure, which negatively affects mechanical performance.

4. Conclusions

This study pioneered TiO_2 sol addition method as an effective approach to prepare Cu–TiO_2 nanocomposite coatings. A greater content of TiO_2 sol upto 12 mL/L resulted in a more refined coating surface with a smaller crystal size. The trend reversed beyond 12 mL/L and a 20 mL/L addition led to a nonuniform surface morphology. The 12 mL/L concentration of TiO_2 sol was recommended for optimal mechanical properties of the coatings.

References

1. P. Y. Guo *et al.*, *Coatings* **8**, 11 (2018).
2. W. Gao *et al.*, *Surf. Coat. Technol.* **350**, 801 (2018).
3. Z. He *et al.*, *J. Electrochem. Soc. D* **165**, 670 (2018).
4. Y. Wang *et al.*, *Surf. Coat. Technol.* **349**, 217 (2018).
5. W. H. Zhang *et al.*, *Int. J. Mod. Phys. B* **33**, 8 (2019).
6. M. Hasegawa *et al.*, *Electrochim. Acta* **178**, 458 (2015).
7. P. T. Lee *et al.*, *Surf. Coat. Technol.* **320**, 559 (2017).
8. M. Lekka *et al.*, *Surf. Coat. Technol.* **205**, 3438 (2011).
9. A. K. Pradhan and S. Das, *Metall. Mater. Trans. A* **45**, 5708 (2014).
10. T. Denny, L. Ronny and B. Andreas, *Electrochim. Acta* **52**, 7362 (2007).
11. Y. Wang *et al.*, *J. Alloy. Compd.* **649**, 222 (2015).
12. M. H. Allahyarzadeh *et al.*, *J. Alloy. Compd.* **705**, 788 (2017).
13. Y. Wang *et al.*, *J. Electrochem. Soc. D* **161**, 775 (2014).
14. Y. Wang *et al.*, *J. Alloy. Compd.* **617**, 472 (2014).
15. Y. Wang *et al.*, *Int. J. Electrochem. Sci.* **9**, 4384 (2014).
16. J. Kong *et al.*, *Electrochim. Acta* **53**, 2048 (2007).

https://doi.org/10.1142/9789819818433_0016

Chapter 16

Laser Cladding of Ni60 + 17-4PH Composite for a Cracking-Free and Corrision Resistive Coating#

Zhaohui Chen*, Ruifeng Li*,§, Jiayang Gu†, Zhongyu Zhang†, Yanwu Tao† and Yingtao Tian‡

*School of Materials Science and Engineering, Jiangsu University of Science and Technology, Zhenjiang 212003, P. R. China

†Marine Equipment and Technology Institute, Jiangsu University of Science and Technology, Zhenjiang 212003, P. R. China

‡Department of Engineering, Lancaster University, Bailrigg, Lancaster LA1 4YW, UK

§li_ruifeng@just.edu.cn

In order to obtain a laser-cladded coating with no cracking and good corrosion resistance, this paper investigated laser cladding of a mixture of 17-4PH stainless steel and Ni60 powders on ASTM 1045 steel substrate. The surface cracking, mechanical properties and corrosion resistance of the coatings were assessed by various characterization methods. The experimental results demonstrated that a crack-free coating can be obtained by adding 30% (or above) 17-4PH stainless steel into Ni60 alloy. The mechanical properties of the coatings were determined by adding 17-4PH, but stabilized at about 79% of pure Ni60 alloy, which is acceptable considering the benefit of elimination of surface cracking. Decrease in the mechanical properties were caused by the dilution of the strengthening elements and reduction of population of hard phases. Composite coating having 30% of 17-4PH also exhibited the smallest corrosion current, lowest corrosion potential and slowest corrosion rate, and therefore the best corrosion resistance.

Keywords: Laser cladding; Ni60; 17-4PH; surface cracking; microhardness; corrosion resistance.

§Corresponding author.

#To cite this article, please refer to its earlier version published in the *International Journal of Modern Physics B*, Volume 34, 2040042 (2020), DOI: 10.1142/S0217979220400421.

1. Introduction

Laser cladding is a surface modification technique that uses laser beam to melt powder materials creating a strong metallurgical bond between the coating and substrate.[1,2] Laser cladding has gradually become an important surface engineering technology and been widely used in surface modification, remanufacturing and repairing in various industrial applications, such as aerospace, oil and gas, power generation and metal processing, which often require high performance of components in harsh environments. For example, steel parts coated with laser-cladded nickel-based alloy often have very high hardness, excellent wear and corrosion resistance, and also high toughness, which can significantly prolong service life and reduce the cost of production.[3,4]

Ni60 is an Ni-Cr-Fe-based alloy offering high hardness, excellent corrosion, wear and heat resistance, which is usually used in thermal spraying process. When it is applied to laser cladding process, the rapid heating and cooling cycle will induce rigorous metallurgical reaction and result in less-favorable microstructures. The γ-Ni supersaturated solid solution can be easily formed and precipitated during multipass laser cladding process. The crystal lattice distortion can induce high level of internal stress in the coating and lead to cracking or pre-matured failure during service.[5,6] In theory, cracks can occur when the internal stress exceeds the yield strength or the local plasticity is insufficient to accommodate the induced plastic deformation, which is often observed in laser cladding of nickel chromium-based alloys.[7] Moreover, the presence of large amount but nonuniformly distributed Cr-rich hard phases can also increase the susceptibility of cracking and lead to high brittleness.[8–10]

17-4PH is a precipitation hardening martensitic stainless steel that offers an outstanding combination of high tensile/impact strength and high fracture toughness. It is also able to withstand relatively high-temperature up to around 300°C without much degradation of overall mechanical properties. Type 17-4PH powder has been widely used for laser cladding and recently in additive manufacturing for nuclear energy aerospace, marine and chemical processing applications.[11–13] Although 17-4PH offers good corrosion resistance in mild acids, it is still far from satisfactory in corrosive demanding environment, as compared with Ni60 alloy.

This study attempts to obtain a high strength, high toughness, crack-free and highly corrosion resistive composite coating by combining the advantages of both Ni60 and 17-4PH alloys. In this paper, mixtures of pre-alloyed Ni60 and 17-4PH powders in various proportions will be used as feedstock for laser cladding on an ASTM 1045 steel substrate. The benefits of adding 17-4PH stainless steel into Ni60 powder will be investigated in terms of surface cracking, changes in microstructure, microhardness and corrosion resistance, etc. The composite powders developed in this study will be promising to solve the cracking problems in conventional Ni60 laser cladding process.

2. Materials and Experiments

The substrate for laser cladding is an ASTM 1045 steel plate in the dimensions of 100 × 100 × 10 mm. The cladding materials are mixtures of 17-4PH stainless steel and Ni60 nickel-based alloy powders, with diameter ranging between 45–106 μm. Four types of powder mixture (Ni60 + 10% 17-4PH, Ni60 + 20% 17-4PH, Ni60 + 30% 17-4PH and Ni60 + 40% 17-4PH) were obtained by using a QM-3SP2 planetary ball mill. Before laser cladding, the powders was heated to 150°C and kept for 30 min in a drying chamber to eliminate the moisture. After drying, the powders were put into a DPSF-2 type coaxial powder feeder for laser cladding experiments.

Laser cladding experiments were conducted by using an IPG high power fiber laser (YLS-6000) at 1600 W output power and a scanning speed of 4 mm/s with a laser spot size of 5 mm. The powder feeding rate was 4.44 g/min assisted by a feeding argon gas flow rate of 12 L/min. The overlap ratio was kept as 30% for all of the experiments. Once the laser cladding was completed, the surface cracks of the cladding layer were characterized by dye penetration tests. The microstructure was observed by using a Zeiss Merlin Compact scanning electron microscope (SEM). The phase in the cladding layer was analyzed by a Shimadzu XRD-6000X X-ray diffraction (XRD) system and the microhardness was tested by a fully automated FM-ARS900 Vickers microhardness tester at 0.2 kg load. Finally, the corrosion resistance was tested by a Coster CS2350 electrochemical workstation.

3. Results and Discussion

In order to quantitatively study the influences of different contents of 17-4PH on the length of surface cracks in the cladding coatings, the length of surface cracks were measured and illustrated in Fig. 1. In the situation where no 17-4PH was added in, the total length of the surface crack was about 47.6 mm. As the 17-4PH powder was blended at a percentage of 10% and 20%, the surface crack length dropped to 23.4 mm and 18.5 mm,

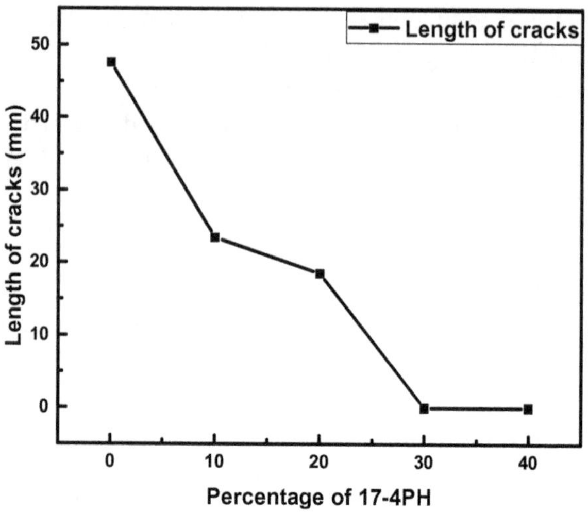

Fig. 1. Total crack length of coatings with different 17-4PH content.

respectively. When the contents of 17-4PH increased to over 30%, the surface crack was hardly visible, so the crack length dropped to virtually zero. Therefore, the dye penetration tests proved that the addition of 30% or more 17-4PH in Ni60 powders can effectively inhibit the surface cracks formation during laser cladding.

The microstructure of laser cladded coatings of various mixtures were observed by using SEM as shown in Fig. 2. The microstructure of pure Ni60 coating has a typical hypoeutectic morphology, which is mainly composed of γ-Ni solid solution dendrites and Cr_7C_3, Ni_5Si_3 and Fe_2B particles among the dendrites.[14] During solidification, the hard phases Cr_7C_3 and Ni_5Si_3 nucleate between the dendrite arms and therefore hinder the coherence of the matrix. The liquid film between the dendrites and hard phases virtually has no ability to sustain any strain induced by the shrinkage resulting in microcracks and porosities. As shown in Fig. 2(a), pure Ni60 alloy has a large number of microcracks within the eutectic phase that can act as a seed for further cracking. After adding 17-4PH in Ni60, the microstructure of the cladded coatings changed significantly. Due to the addition of 17-4PH powder, the content of C, Si and B elements was diluted which reduces the driving force for the diffusion-controlled formation of the abovementioned hard phases. When the tip of the γ-Ni dendrite reaches the hard particles, the growth is then interrupted so that the continuity

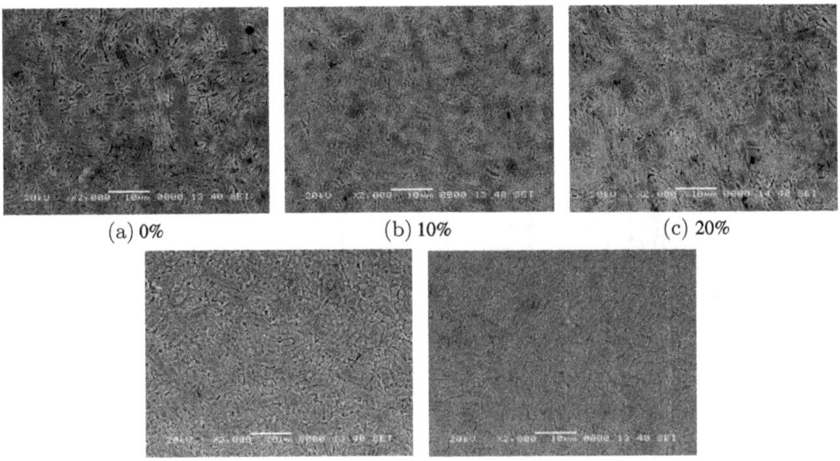

(a) 0% (b) 10% (c) 20% (d) 30% (e) 40%

Fig. 2. Microstructure of different 17-4PH cladding layer added to Ni60.

of the matrix is deteriorated.[15,16] As illustrated in Figs. 2(b)–2(d), when the content of 17-4PH increased from 10% to 30%, the population of microcracks gradually reduced. Although no macroscale surface crack was observed in the case of 30% 17-4PH, as shown in Fig. 1, cracks in micrometer scale can still be spotted under high magnification of SEM [Fig. 3(d)]. When the content of 17-4PH increased to 40%, the population of the hard phases was significantly reduced and the γ-Ni solid solution and eutectic substrate became more coherent and almost crack-free, as shown in Fig. 2(e).

Figure 3 compares the XRD patterns of pure Ni60 laser cladded coatings and with 30% of 17-4PH. As discussed above, solidification during laser cladding process is highly nonequilibrium and intrinsically leads to supersaturation in the matrix and lattice distortion. The XRD pattern shown in Fig. 3(a) confirms that the pure Ni60 alloy coating mainly consists of γ-Ni solid solution and hard phases such as Cr_7C_3, Ni_5Si_3 and Fe_2B. With 30% of additional 17-4PH, these hard phases disappear, instead, the composite coating mainly consists of γ-Ni, α-Fe and γ-Fe, as shown in Fig. 3(b).

Figure 4 plots the microhardness on top surface/center/bottom of the coatings, heat affected zones and substrates. An averaged microhardness of 575 $HV_{0.2}$ was obtained in pure Ni60 coating and the microhardness distribution was almost identical when only 10% of 17-4PH powder was

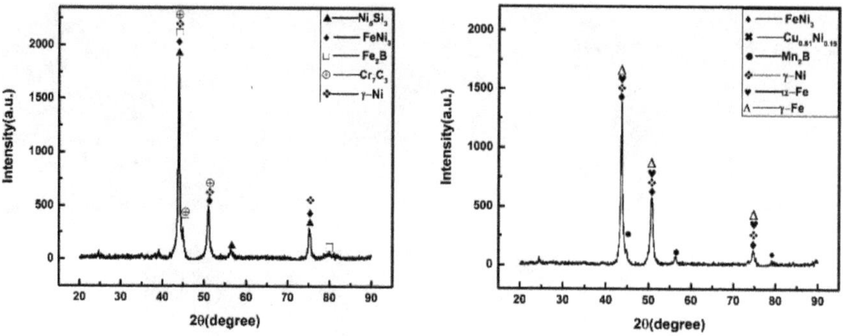

(a) 0% (b) 30%

Fig. 3. XRD pattern of different 17-4PH content cladding layer.

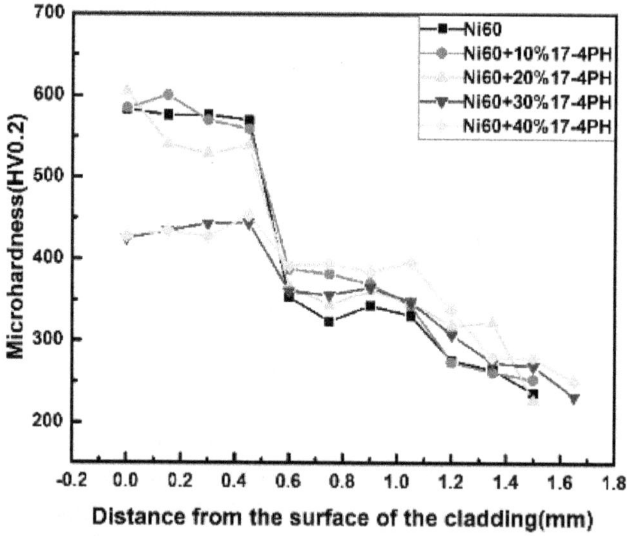

Fig. 4. (Color online) Hardness distribution of different 17-4PH content cladding layers.

added. When 20% 17-4PH was blended in, the averaged microhardness of the cladded coating slightly decreased, i.e., to about 550 $HV_{0.2}$. microhardness demonstrates a dramatic decrease with the addition of 30% and 40% 17-4PH, to only about 450 $HV_{0.2}$ which is approximately 21% less than pure Ni60, but the difference between the microhardness of coatings having 30% and 40% of 17-4PH was trivial. The heat-affected zone (HAZ) was actually part of the substrate before laser cladding, due to the rapid heating

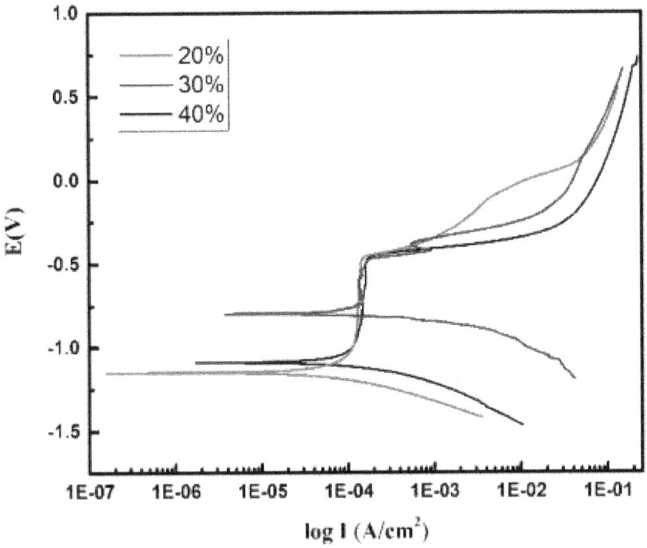

Fig. 5. (Color online) Electrochemical corrosion polarization curves of different 17-4PH content cladding layers.

and cooling during laser cladding process, the HAZ region undergoes an equivalent heat treatment and therefore exhibits a microhardness higher than the original substrate. In all cases, the microhardness of HAZ was about 100 $HV_{0.2}$ higher than that of the substrate.

The corrosion resistance of the composite coatings was evaluated by running a number of polarization curves in 3.5%NaCl solution. Figure 5 plots the polarization curves of laser cladded composites containing 20%, 30% and 40% of 17-4PH, respectively. When the potential increases, a dynamic passivation layer is formed until a so-called breakdown potential (E_b) is reached. Then, the passivation film begins to dissolve and a surge of current can be observed. It is inaccurate to assess the corrosion resistance on the basis of the breakdown potential because it can be affected by the self-corrosion potential. The polarization curves can be interpreted by using the C-View software and the corresponding corrosion current, breakdown potential and corrosion rate can be thereafter obtained. It is found that the corrosion current I_{corr} is the smallest when the content of 17-4PH is 30% and the corrosion potential E_{corr} also moves closer to the neutral position. A corrosion rate of 0.33594 (mm/a) was also the smallest among the assessed three situations. The reason for a higher corrosion rate occurring in 40%

of 17-4PH composite is probably due to the insufficient γ-Ni solid solution and the increased fraction of α-Fe.[17]

4. Conclusions

In this study, laser cladding of composite coatings by blending 17-4PH stainless steel into Ni60 powders were investigated and the following conclusions can be drawn based on the experimental results:

(1) Adding 17-4PH stainless steel into Ni60 powder can effectively inhibit the formation of the coarse and brittle hard phase in the coating and suppress the surface cracks formation. A crack-free coating can be obtained when the content of 17-4PH reaches 30% or above.
(2) The mechanical strength of the composite coatings decreases about 21% of pure Ni60 while the content of 17-4PH increase to over 30%. The drop in the mechanical strength is justified by the elimination of surface cracking.
(3) Composite coating consisting of 30% 17-4PH exhibits the smallest corrosion current, the lowest corrosion potential and corrosion rate and therefore the best corrosion resistance.

Acknowledgments

The authors would like to acknowledge the financial support provided by the National Key Research and Development Program of China (No. 2018YFC0310400), the National Natural Science Foundation of China (Grant No. 51775254).

References

1. M. Yue *et al.*, *Surf. Interface Anal.* **29**, 310 (2010).
2. W. Fei, C. Chen and H. Yu, *Mater. Des.* **58**, 412 (2014).
3. F. Fu *et al.*, *Optik – Int. J. Light Electron Opt.* **127**, 200 (2012).
4. X. Luo, J. Li and G. J. Li, *J. Alloy. Compd.* **626**, 102 (2015).
5. J. Zhang *et al.*, *Trans. Nonferr. Metals Soc. China* **25**, 1525 (2015).
6. A. N. Grezev and A. N. Safonov, *Weld. Int.* **22**, 50 (1987).
7. Y. U. Ting *et al.*, *J. Shanghai Jiaotong Univ.* **46**, 1043 (2012).
8. Y. U. Ting and Deng, *J. Mech. Eng.* **47**, 25 (2011).
9. C. Z. Chen and T. Q. H. Lei, *Mater. Sci. Technol.* **10**, 431 (2002).
10. J. Wang *et al.*, *Mater. Charact.* **57**, 274 (2006).
11. R. Rashid *et al.*, *J. Mater. Process. Technol.* **249**, 502 (2017).
12. Z. J. Liu, X. M. Liu and X. Li, *Power Eng.* **23**, 2201 (2003).

13. C. Tassin *et al.*, *Surf. Coat. Technol.* **80**, 207 (1996).
14. J. H. Yao *et al.*, *Mater. Des.* **83**, 26 (2015).
15. L. Chen *et al.*, *Chin. J. Lasers* **11**, 69 (2014).
16. W. J. Chen *et al.*, *Eng. Failure Anal.* **79**, 120 (2017).
17. H. Liu *et al.*, *Surf. Coat. Technol.* **228**, S296 (2013).

https://doi.org/10.1142/9789819818433_0017

Chapter 17

Effect of Mg Plate Pretreatment on Microstructure and Mechanical Properties of Al/Mg/Al Laminated Composites Fabricated by Hot Roll Bonding[#]

Yang Yu[*,†,§], Xinzhe Xue[†], Zhehao Lu[†], Tingyu Li[†], Cong Ma[*,†], Pengfei Yan[‡] and Biao Yan[*,†,¶]

[*]*Shanghai Key Laboratory for R&D and Application of Metallic Functional Materials, Shanghai 201804, P. R. China*

[†]*School of Materials Science and Engineering, Tongji University, Shanghai 201804, P. R. China*

[‡]*School of Physics Science and Engineering, Tongji University, Shanghai 200092, P. R. China*

[§]*yuyangmr@163.com*

[¶]*yan_biao@tongji.edu.cn*

The AA7075/AZ31/AA7075 laminated composites were fabricated by hot roll bonding. The effects of pretreatment of Mg plate by severe plastic deformation (SPD) and heat treatment on its interface, microstructure and mechanical properties were studied. The diffusion layer is formed at the interface between Al and Mg layers, which is composed of the hard and brittle intermetallic phases β-Al_3Mg_2 and γ-$Al_{12}Mg_{17}$, and become the weak zone of the laminated composites where the crack is first-generated and propagated to cause the delamination at the interface of the laminated composites. The hardness of laminated composites at the interface reaches 300–400 HV. Meanwhile, Mg plate pretreatment can increase the plastic matching and the deformation compatibility of the Al and Mg layers, and can effectively improve the mechanical properties of the laminated composites to a certain extent. The bending strength of the laminated composites with pretreated Mg plates are 1080 MPa.

Keywords: Al/Mg/Al laminated composites; hot roll bonding; diffusion layer; bending properties.

[¶]Corresponding author.

[#]To cite this article, please refer to its earlier version published in the *International Journal of Modern Physics B*, Volume 34, 2040052 (2020), DOI: 10.1142/S0217979220400524.

1. Introduction

Mg alloy is considered as a potential material for automobile structural components due to its light weight, environmental protection and good damping ability.[1–3] However, there are still two main problems of corrosion and formability in the preparation and application of Mg alloy, which have immensely restricted its application in the industry.[3–5] Al alloy usually has excellent corrosion resistance and formability.[5–7] It is expected to enhance corrosion resistance and mechanical properties by fabricating Al/Mg/Al composite plates, which do not add much weight.[7–9] Therefore, many researches on the preparation, interface, microstructure and mechanical properties of the Al/Mg composites have been reported in the literature. The diffusion layer can be formed at the interface of the Al/Mg composites prepared by methods such as extrusion,[6–8] rolling[2,3] and explosive welding,[10] which is mainly composed of intermetallic phases $Al_{12}Mg_{17}$ and Al_3Mg_2.[2–4] The diffusion layer is mainly related to the subsequent annealing temperature and time; and it is generally believed that with the increase of diffusion layer thickness, the interfacial bonding strength decreases.[1,2]

The edge cracks and surface microcrack can easily form in the Mg layer of the Al/Mg laminated composite during rolling process, mainly due to the poor formability of the magnesium alloy.[11] Thus, developing Al/Mg laminated composites is beneficial to improve the plasticity of Mg plate. Refinement of grain size and weakening of basal texture are considered two crucial ways for improving the formability of magnesium alloy during hot rolling,[12] which can be achieved by severe plastic deformation (SPD)[13] and recrystallization.[14] However, there are few reports on the preparation of Al/Mg composite plates with plastically treated magnesium plates, and the effect of pretreated Mg plates on the microstructure and mechanical properties of the Al/Mg composite plates is yet unclear.

In this work, the Al/Mg/Al laminated composites with pretreatment of Mg plate by SPD were fabricated by hot roll bonding. The effect of pretreated Mg plates on the interface, microstructure and mechanical properties of the Al/Mg/Al laminated composites was investigated. It was found that Mg plate pretreatment can increase the plastic matching and the deformation compatibility of the Al and Mg layers, and improve the mechanical properties of the laminated composites to a certain extent.

2. Experimental Procedures

2.1. *Materials and fabrication process*

Al alloy AA7075-T6 and Mg alloy AZ31B in the as-received and homogenized condition were used for the fabrication of the Al/Mg/Al laminated composites. As-received Al and Mg plates with dimensions of $100 \times 50 \times 2$ mm^3 are for stacking and hot roll bonding. In addition, there are 5 mm thick Mg plates, which are pretreated and rolled to a thickness of 2 mm to improve the plasticity of the Mg plate and compounded with Al plates to form the laminated composites.

Figure 1 shows the schematic diagram of fabrication process of Al/Mg/Al composite plate. The oxide film of the Al and Mg plates is first removed by grinding to obtain fresh surface. Then 2 mm-thick Al and Mg plates are stacked alternately and staggered in sequence of Al–Mg–Al. The four corners of stacked plate are drilled and fixed with aluminum rivets. The fixed plates were heated for 10 min at 400°C in muffle furnace, then rolled twice to obtain 2 mm-thick Al/Mg/Al laminated composites. The total reduction of Al/Mg/Al laminated composites is 66.7%. The rolled laminated composites subsequently annealed at 200°C for 2 h. The same rolling parameters are used for the 5 mm-thick Mg plates to obtain the 2 mm-thick pretreated Mg plates. The total reduction of pretreated Mg plates was 60%.

2.2. *Microstructure characterization and mechanical properties*

The specimens for microstructure and interface analysis were prepared along the rolling direction. Microstructures were observed by a MJ41 optical microscope (OM). Interface morphology was characterized by using

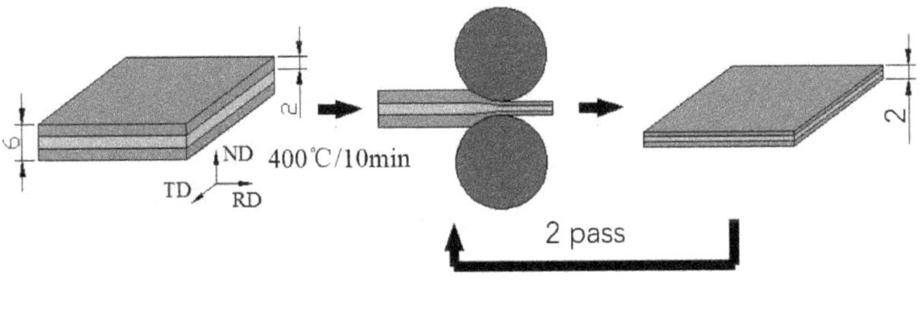

Fig. 1. The schematic diagram of fabrication process of Al–Mg–Al composite plate.

a SU1510 scanning electron microscope (SEM) equipped with the iXRF energy dispersive spectrometry (EDS) system, and intermetallic phases were analyzed by DX-2000 X-ray diffraction (XRD).

Microhardness of specimens was measured on their cross-sections using a HVT-1000A microhardness tester. The loading force was 4.9 kN and the distance between each measuring point was 0.05 mm. Three-point bending tests with initial strain rate of 1×10^{-3} were carried out by a SANS CHT4305 electron universal testing machine.

3. Results and Discussion

3.1. *The interface and microstructure*

Figure 2 shows the interface morphology of the laminated composites without and with pretreated Mg plate. From the interface of the laminated composites without pretreated Mg plate in Fig. 2(a), it can be found that the interface between Al and Mg layers has a diffusion layer. This layer is not smooth and has local triangular protrusions. In contrast, the interface in the specimen made from pretreated Mg plate was relatively smooth, as shown in Fig. 2(b).

The thickness of diffusion layer is mainly related to heat treatment temperature and time. Chen *et al.*[1] suggested that as the annealing temperature increases, the diffusion phenomenon becomes more and more obvious, and the diffusion layer also increases in Mg/Al composite plates. Figure 3 shows the XRD spectrum of cross-section of the laminated composites. The diffusion layer mainly contains intermetallic phases Al_3Mg_2 and $Al_{12}Mg_{17}$. Zhao *et al.*[4] observed intermetallic phases β-Al_3Mg_2 and γ-$Al_{12}Mg_{17}$ at the interface of Al/Mg composite materials by TEM.

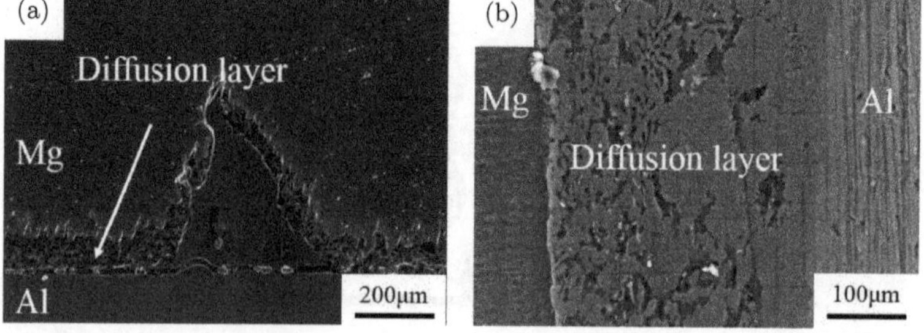

Fig. 2. The schematic diagram of fabrication process of Al–Mg–Al composite plate.

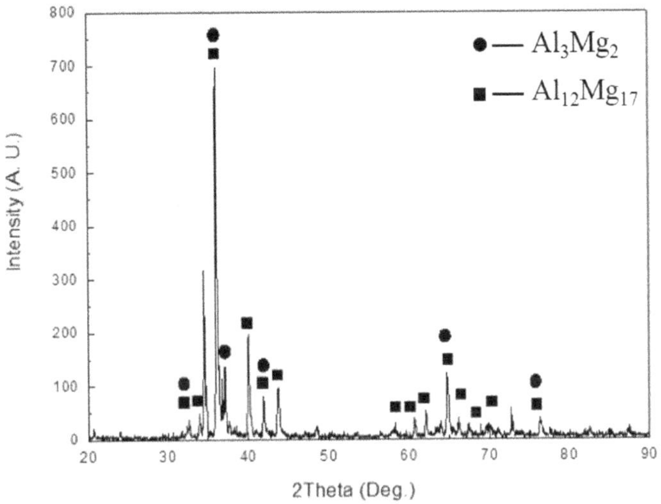

Fig. 3. XRD spectrum of cross-section of the laminated composites.

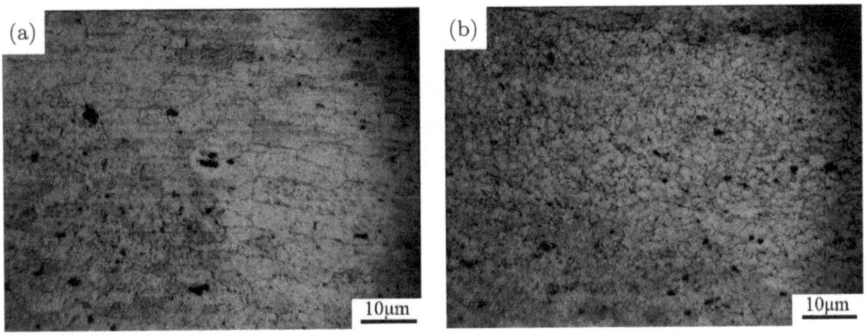

Fig. 4. The microstructure of the (a) as-received Al plate and (b) Al layer in the laminated composite.

Figure 4 shows the microstructure of the as-received Al plate and Al layer of the laminated composite. The microstructure of the as-received Al alloy is coarser, as shown in Fig. 4(a). It can be found in Fig. 4(b) that the grains of Al alloy were refined under the effect of severe plastic deformation and recrystallization after hot roll bonding, with their grain sizes reduced significantly from approximately 10 μm to about 1 μm. Besides, Fig. 5 shows the microstructure of the as-received Mg plate and Mg layer of the

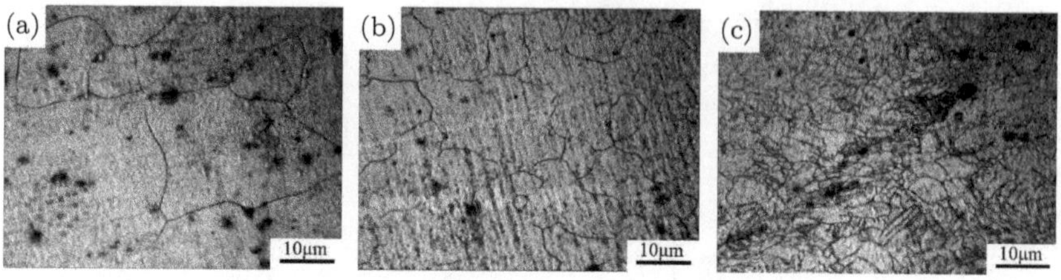

Fig. 5. The microstructure of the (a) as-received Mg plate and Mg layer in the laminated composites (b) without and (c) with pretreated Mg plate.

laminated composite without and with pretreated Mg plate. Comparing Figs. 5(a) and 5(b), it can be found that the grains of Mg alloy were refined as well after hot roll bonding, and the grain size was changed from 30 μm to 15 μm. However, in the laminated composite from pretreated Mg, it is noted in Fig. 5(c) that the grains were much finer than that of the composite from Mg without pretreatment, with grain size of about 5 μm.

Zhi *et al.*[11] reported that the edge cracks and surface microcracks are easy to form in the rolling process of Mg plates. However, in our work, the interface of the laminated composites with pretreated Mg plates is smooth. There is no microcrack on the Mg plate, as illustrated in Fig. 2(b). This further proves that pretreatment on Mg plate contributes to the grain refinement of Mg layer and improves the formability and mechanical properties of the laminated composites which will be discussed in the following.

3.2. *The microhardness and bending property*

Figure 6(a) shows the microhardness distribution contours of cross-section of the Al/Mg/Al laminated composites without pretreated Mg plates. It can be found that the hardness distribution inside the Al and Mg layers is relatively uniform. The average hardness of the Al layers is 100–200 HV, and the hardness of the Mg layer is the lowest, about 60–80 HV. However, the hardness distribution of the diffusion layer at the interface is complex and uneven, and its hardness is the highest, about 300–400 HV.

Figure 6(b) shows the bending curves of the Al/Mg/Al laminated composites without and with pretreated Mg plates. It can be found from the bending curves that the two bending curves are completely different. Among them, there are turning points in the bending curve of the laminated composites without pretreated Mg plates, while the bending curve with

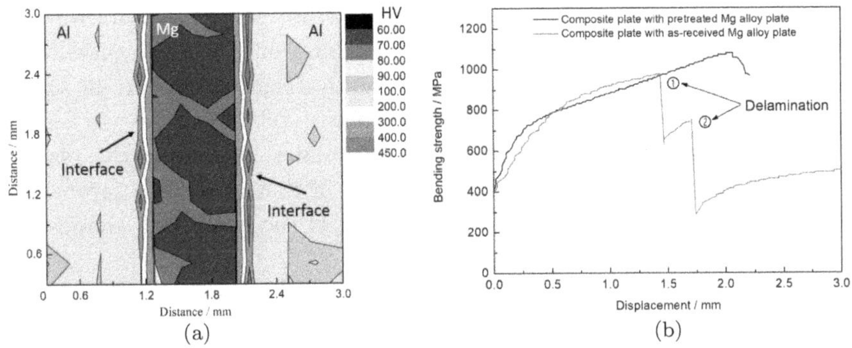

Fig. 6. (Color online) (a) Microhardness distribution contours of cross-section of the Al/Mg/Al laminated composites without pretreated Mg plates and (b) the bending curves of the Al/Mg/Al laminated composites without and with pretreated Mg plates.

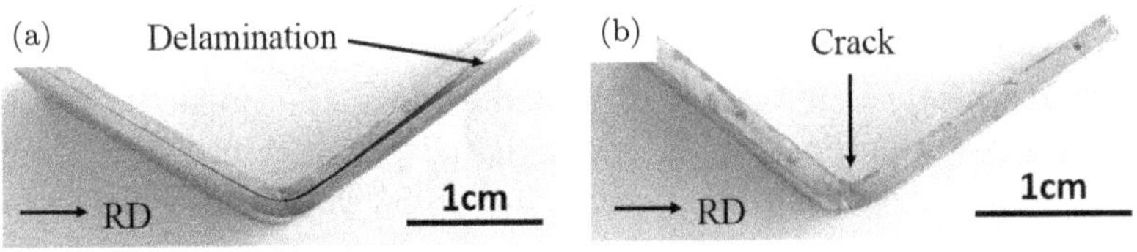

Fig. 7. (Color online) The macroscopic fracture morphology of bending specimens of the Al/Mg/Al laminated composites (a) without and (b) with pretreated Mg plates, RD: rolling direction.

pretreated Mg plates has no turning point. It can be found that the bending strength with pretreated Mg plates is the highest, reaching 1080 MPa.

Figure 7 shows the macroscopic fracture morphology of bending specimens of the Al/Mg/Al laminated composites without and with pretreated Mg plates. It can be found in Fig. 7(a) the delamination phenomenon in the laminated composites without pretreated Mg plates. Feng *et al.*[6,7] and Wu *et al.*[8] studied many extruded Al/Mg composites and found that cracks first appeared in the diffusion layer during loading and deformation. This indicates that the diffusion layers are the weak zones of the laminated composites. However, there is no delamination in the laminated composites with pretreated Mg plate but a crack is formed in the outer Al layer with the most serious bending deformation, as shown in Fig. 7(b).

The laminated composites with pretreated Mg plates have better strength and plasticity than those with pretreated Mg plates, which indicates that the plasticity improvement of Mg layer can improve the deformation compatibility of the laminated composites to a certain extent. Feng *et al.*[6] also mentioned this good deformation compatibility in the study of Mg AZ31 core/hard Al 7050 sleeve composite rods. Therefore, as brittle phases form at the interface of Al and Mg plates, increasing the plastic matching and the deformation compatibility of the Al and Mg layers can effectively improve the mechanical properties of the laminated composites to some extent.

4. Conclusions

(1) The diffusion layer is formed at the interface between Al and Mg layers, which is composed of intermetallic phases β-Al_3Mg_2 and γ-$Al_{12}Mg_{17}$. These two intermetallic phases belong to the hard and brittle phase, which makes the hardness of laminated composites at the interface higher, reaching 300–400 HV. The diffusion layer is the weak zone of the laminated composites.

(2) Pretreatment of Mg plate by SPD and heat treatment can refine grains of Mg layer, and improve its strength and plasticity of Mg plates and the laminated composites. The bending strength of the laminated composites with pretreated Mg plates is 1080 MPa. Meanwhile, Mg plate pretreatment can increase the plastic matching and the deformation compatibility of the Al and Mg layers. It can also effectively improve the mechanical properties of the laminated composites to a certain extent.

Acknowledgment

This research was supported by the National Key Research and Development Program of the 13th Five-Year Plan of China (Grant No. 2016YFB1200602-02).

References

1. Z. Q. Chen *et al.*, *Mater. Sci. Eng. A-Struct.* **723**, 97 (2018).
2. H. P. Zheng *et al.*, *Adv. Eng. Mater.* **18**, 1792 (2016).
3. X. P. Zhang *et al.*, *Mater. Sci. Eng. A-Struct.* **528**, 1954 (2011).
4. Y. D. Zhao *et al.*, *Mater. Charact.* **128**, 156 (2017).
5. Q. H. Wang *et al.*, *Mater. Sci. Eng. A-Struct.* **731**, 184 (2018).
6. B. Feng *et al.*, *Acta Materialia.* **120**, 379 (2016).
7. B. Feng *et al.*, *Mater. Sci. Eng. A-Struct.* **675**, 204 (2016).

8. Y. Wu *et al.*, *Mater. Sci. Eng. A-Struct.* **640**, 454 (2015).
9. C. Y. Liu *et al.*, *Mater. Sci. Eng. A-Struct.* **556**, 1 (2012).
10. Y. B. Yan *et al.*, *Chin. J. Nonferrous Met.* **20**, 674 (2010).
11. C. C. Zhi *et al.*, *J. Mater. Process Technol.* **255**, 333 (2018).
12. W. Z. Chen *et al.*, *Scripta Mater.* **67**, 858 (2012).
13. M. T. Pérez-Prado *et al.*, *Mater. Lett.* **59**, 3299 (2005).
14. Q. Xu *et al.*, *Mater. Lett.* **64**, 1085 (2010).

https://doi.org/10.1142/9789819818433_0018

Chapter 18

Study of Parameters Affecting Microcontact Printing of Thiols on Gold-Coated Substrate#

Swapna A. Jaywant and Khalid Mahmood Arif*

School of Food and Advanced Technology,
Massey University, 229 Dairy Flat Highway,
Albany, Auckland 0632, New Zealand
**k.arif@massey.ac.nz*

Microcontact printing (μCP) is a type of soft-lithography technique, which is widely used for patterning self-assembled monolayers (SAMs). It is a convenient method to form SAMs of bio/chemical ink onto different surfaces such as polymers, palladium, silver and gold. A wide range of applications of this technology includes micromachining, patterning proteins, cells or DNA in biosensors. However, the application primarily depends on the type of the ink used. Here, we present an experimental study that provides information about the parameters that affect the μCP process. Two different thiol inks (dithiothreitol (DTT) and glutathione (GSH)) have been used for obtaining SAMs on gold-coated substrates. Our findings suggest that transferring the alkanethiols over the gold surface is extremely dependent upon the molecular weight of thiol compound, concentration of the thiol solution and pH value of the buffer used. Furthermore, higher the molecular weight, concentration and pH value of the ink, lower is the time required for the process of μCP.

Keywords: Microcontact printing; self-assembled monolayers; thiol compound.

1. Introduction

Microcontact printing (μCP) is gaining popularity in the research areas such as BioMEMS, Lab on a Chip (LoC) devices or cell biology for surface passivation using patterned monolayers with submicron resolution.[1] This technique was introduced by Whitesides and his colleagues in 1993.[2]

*Corresponding author.

#To cite this article, please refer to its earlier version published in the *International Journal of Modern Physics B*, Volume 34, 2040040 (2020), DOI: 10.1142/S0217979220400408.

It is comparable to the conventional stamping method as it comprises of an ink, a stamp and a substrate. A polymeric stamp with respite pattern is generally used in μCP to produce the micro- or nano-scale patterns. The stamp is linked by either putting a drop of ink (proteins, etc.) on the stamp or immersing the stamp in the ink. Thereafter, the stamp is brought into contact with the substrate surface. Polydimethylsiloxane (PDMS) is a widely used material to make the stamps for μCP.[3]

The application of this method is primarily dependent on the type of the ink used. Therefore, a variety of inks are used like thiol compounds, synthetic polymers, dendrimers, proteins, DNA, etc. Hence, different applications of μCP are available such as micromachining, pattering proteins, cells or DNA in biosensors on various surfaces such as gold, silver, copper, glass, polymers, etc.[4] Patterning self-assembled monolayer (SAM) of alkane thiolates or thiols using μCP is currently considered as a valuable application due to its superior control over the surface chemistry and parallel processing.[5] SAM consists of a head domain that binds with the substrate surface, a tail which forms ordered structures away from the substrate surface Fig. 1(a). Generally, tails have a functional end group. Most commonly, thiol molecules are used to form the SAMs. Thiol is an organic compound that comprises a carbon-bonded sulfhydryl (–SH) group. In the case of thiols, sulfur atom covalently binds with the surface of the substrate and terminal functional group changes the surface chemistry of the substrate.[6] Several advantages of μCP have been reported in the literature, e.g., (i) it is not as expensive as photolithography technique, (ii) it can create multifunctional nano/microstructures on various surfaces and (iii) usage of the cleanroom is not mandatory.[3,7] Despite all the

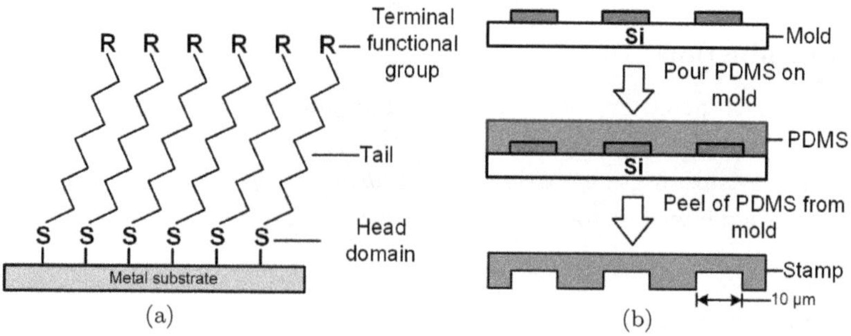

Fig. 1. (Color online) (a) Structure of self-assembled monolayer and (b) fabrication of PDMS stamp.

advantages of this method, there are a few limitations too, such as stamp deformation,[3,7] volatility of ink and diffusion of the ink molecules due to longer contact time. Furthermore, water-soluble ink does not permeate due to hydrophobic surface properties of PDMS which affects the printed SAM quality.[3,8] Through the experiments, we also found that there are additional factors which affect μCP process such as molecular weight and pH value of the ink.

We explored two different thiol inks for obtaining SAMs. One of the thiols is dithiothreitol (DTT), which is a dithiol compound with two thiol functional groups in it. The other thiol is the glutathione (GSH), which is a monothiol compound having one thiol functional group in it. The molecular weight of DTT is lower compared with GSH. We used gold-coated substrates during the μCP process as it is a rationally inert material, hence samples can be easily handled under atmospheric conditions. Gold also has a very high affinity towards the thiols.[9–11]

2. Experimental Methods

Analytical reagent-grade chemicals were used. All chemicals were purchased from Sigma Aldrich, New Zealand. GSH ($C_{10}H_{17}N_3O_6S$), DTT ($C_4H_{10}O_2S_2$), phosphate buffer saline (PBS pH 7.4), pH 10 buffer solution and ethanol were used as received without modification. Deionized (DI) water was used for all experiments. SYLGARD® 184 silicone elastomer kit was also purchased from Sigma Aldrich, NZ. Glass slides were purchased from Thermo Scientific, NZ. Desk Sputter Coater (DSR1) was used to prepare the gold-coated glass substrate. An optical microscope (Nikon-H600L) was used to observe the self-assembled patterns.

2.1. *Substrate and stamp fabrication*

The gold-coated glass substrate was prepared in the laboratory. Here, the glass slides were cut into the pieces of size 0.8 cm × 0.8 cm. Next, these pieces were cleaned with ethanol followed by rinsing in DI water and drying with an ultrahigh purity nitrogen (BOC, NZ) stream prior to sputtering. The layer of gold (50 nm thickness) was sputtered on these glass slides using the sputtering machine DSR1. These gold-coated substrates were cleaned with DI water and dried with nitrogen gas before use. A new set of gold-coated substrates was prepared separately for each experiment.

To fabricate the stamps, a master mold having 10 μm alternating patterns was used. The stamp master was obtained from BW Foundry,

The University of Sydney. During fabrication of the stamps, the silicone elastomer base was mixed with a curing agent in a 10:1 ratio, degassed for 1 h and then poured onto the master mold.[12,13] Later, at 60°C, curing was carried out for 3 h which was followed by cooling at room-temperature. The cured stamp was gradually peeled off the mold and carefully stored in a cleaned petri dish. Figure 1(b) pictorially depicts the stamp fabrication process.

2.2. *Inking and printing*

Inking of the PDMS stamp is an important step during μCP process. There are various well-known inking methods: liquid inking, inking through PDMS reservoir, direct inking and inking by immersing PDMS stamp in the ink.[4,12] We follow the immersion inking method in our experiments. The PDMS stamp was cut into smaller pieces (0.5 cm $\times$ 0.5 cm) before stamping. The smaller stamps were cleaned using an ultrasonic cleaner with ethanol for half an hour at 30°C before use. Different concentrations of GSH and DTT were prepared in two different buffers (PBS and buffer solution with pH 10). The stamp was immersed in the prepared solution for inking purpose.

To transfer the ink on the substrate, printing was performed. For this purpose, PDMS stamp was removed from the thiol ink. Any excess thiol solution was removed from the stamp using a moderate stream of nitrogen gas. Next, the stamp was brought into contact with the gold-coated substrate. Gentle pressure was applied on the stamp to ensure the precise contact between the gold coating and stamp. The stamp was removed from the substrate, which was followed by rinsing the substrate using DI water to remove any unbounded thiol molecules.

3. Results and Discussion

Consistent patterns of DTT and GSH compounds were achieved by stabilizing the thiol molecules on the gold-coated substrate through Au—S linkages.[11,14–16] Figures 2(a) and 2(b) illustrate the interaction of gold with thiol molecules and Fig. 2(c) shows the obtained patterns with DTT and GSH. The strength of the Au—S interaction formed on the gold surface depends upon various parameters such as the type of buffer, inking period, concentration of the thiol solution, etc. The inking period is the duration of the immersion of a PDMS stamp in the thiol solution. Hence, patterns with different thiol concentrations, inking periods and pH values were formed

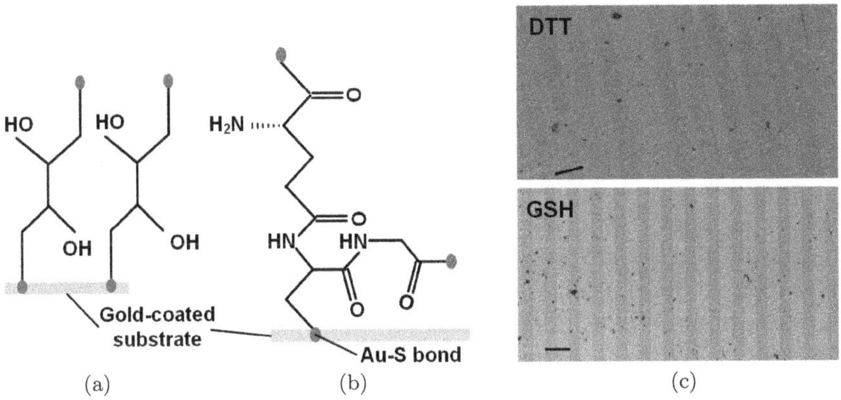

Fig. 2. (Color online) Interaction of gold with (a) DTT, (b) GSH molecules and (c) optical patterns obtained with DTT and GSH on a gold-coated substrate. Scale bar: 10 μm.

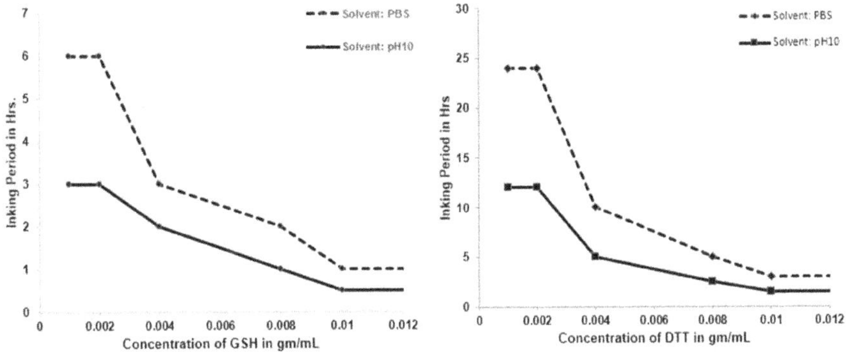

Fig. 3. Effect of concentration and pH value on inking period.

for characterization purpose. To observe the effect of concentration on the patterns, different concentrations of GSH and DTT (0.002, 0.004, 0.008 and 0.010 gm/mL) solutions were prepared. Two buffers with different pH values were used to prepare the concentrations so that the effect of pH value on the pattern could be observed. To study the effect of inking period, the PDMS stamp was immersed in each solution for different time periods (24, 12, 6, 5, 3, 2, 1 and 0.5 h) and then brought into contact with the gold-coated substrate. This was followed by observing the patterns under the optical microscope. The experimental results are plotted in Fig. 3.

It was found that to obtain a visible pattern, the minimum concentration of GSH should be at least 1 mg/mL. Further investigations discovered that for the concentrations upto 2 mg/mL using PBS as a buffer, inking period must be at least 6 h to obtain the pattern. In the case of pH 10 as a buffer, inking period had to be 3 h. Furthermore, as GSH concentration increases, there is a corresponding decrease in inking period for both the buffers. However, minimum 1 h is required to obtain the pattern from 10 mg/mL and above when GSH is diluted with PBS whereas, minimum half an hour is enough using pH 10 as a buffer. In the case of DTT, it has been discovered that for the concentrations upto 2 mg/mL using PBS as a buffer, inking period must be at least 24 h to obtain the pattern. In case of pH 10 as a buffer, inking period must be 12 h. Similar to GSH, as DTT concentration rises, there is a consistent reduction in inking period for both the buffers. However, a minimum of 3 h are essential to obtain the pattern from 10 mg/mL and above using PBS whereas, minimum an hour is adequate using pH 10 as a buffer.

The effect of pH value on producing the micropatterns was verified with the help of two different buffers i.e., pH 7.4 and pH 10. The plots in Fig. 3 show a comparison between the experimental results of obtained micropatterns using different pH values. It is clearly indicated that the required inking period decreases as there is an increase in pH value. It is also observed that the usage of pH 10 buffer requires almost half the value of inking period than PBS.

The molecular weight of thiol compound plays a key role while transferring the ink on the gold substrate. Monothiol GSH molecule is nearly two times heavier than the dithiol DTT molecule. The molecular weight of DTT is 154.253 g/mol whereas GSH has 307.325 g/mol molecular weight. The experimentation result implies that patterns of GSH molecules take less time to be transferred on the gold-coated substrate when compared with DTT molecules. For example, it took 6 h to transfer the pattern successfully with 0.002 gm/mL concentration in PBS solution for GSH molecule. On the other hand, for DTT molecule with the same concentration and buffer, 24 h were required for the pattern transfer.

4. Conclusion

In conclusion, μCP is a very useful technique in patterning the SAMs of bio/chemical materials. However, the effectiveness of the technique depends upon various parameters such as the type of ink, buffer used to prepare the ink, concentration of the ink, inking period of the PDMS stamp, pH value of the buffer and molecular weight of the ink material. In our study, we used thiol ink during the process of μCP due to the popularity of thiols in numerous applications. The study indicates that the total time taken by the μCP process to transfer the patterns successfully on the gold-coated substrate is inversely proportional to the concentration of the thiol molecules and pH value of the buffer. When pH value changes from 7.4 to 10, there is a significant reduction in the essential inking period. Lastly, the success of pattern transferring using μCP is also dependent on the molecular weight of the thiol molecule. Higher the molecular weight, lower the inking period and so lower is the total length of time taken by μCP process.

Acknowledgments

We thank Massey University, School of Food and Advanced Technology for providing funds for this work.

References

1. A. Agarwal *et al.*, *Lab Chip* **13**, 3599 (2013), doi:10.1039/C3LC50350J.
2. J. L. Wilbur *et al.*, *Adv. Mater.* **6**, 600 (1994).
3. A. Perl, D. N. Reinhoudt and J. Huskens, *Adv. Mater.* **21**, 2257 (2009).
4. L. Libioulle *et al.*, *Langmuir* **15**, 300 (1999).
5. Y. Xia, X.-M. Zhao and G. M. Whitesides, *Microelectron. Eng.* **32**, 255 (1996).
6. M. Boeckl and D. Graham, *Mater. Matters* **1**, 3 (2006).
7. C. Y. Hui *et al.*, *Langmuir* **18**, 1394 (2002), doi:10.1021/la0113567.
8. F. Bessueille *et al.*, *Langmuir* **21**, 12060 (2005).
9. C. D. Bain, J. Evall and G. M. Whitesides, *J. Am. Chem. Soc.* **111**, 7155 (1989).
10. J. C. Love *et al.*, *Chem. Rev.* **105**, 1103 (2005).
11. Y. Xue *et al.*, *Nat. Commun.* **5**, 4348 (2014).
12. T. Kaufmann and B. J. Ravoo, *Polym. Chem.* **1**, 371 (2010).
13. S. A. Ruiz and C. S. Chen, *Soft Matter* **3**, 168 (2007).
14. A. MacDairmid, M. Gallagher and J. Banks, *J. Phys. Chem. B* **107**, 9789 (2003).
15. D.-H. Tsai *et al.*, *Langmuir* **30**, 3397 (2014).
16. P. D. Jadzinsky *et al.*, *Science* **318**, 430 (2007).

https://doi.org/10.1142/9789819818433_0019

Chapter 19

Effect of Graphene Lateral Size on the Microstructure and Properties of Graphene/Copper Composites#

Dandan Zhang*, Huaxing Xiao, Wei Jiang, Xia Cao,
Manyu Ye, Lu Huang and Tao Tang

Department of Material Shaping and Control Technology,
Changzhou Institute of Technology,
Changzhou 213032, P. R. China
**zhangdd@czust.edu.cn*

Copper matrix composites reinforced with graphene nanoplatelets (GNPs) were prepared by vacuum hot pressing of ball milled mixtures of powders. Two grades of GNPs were used; one with average thickness of 2 nm and average lateral size of 6 μm and another with much larger lateral size of 80 μm. Microstructure and properties of as-prepared composites containing 10 vol.% GNPs were studied. The GNPs sheets are uniformly distributed and well aligned in the Cu matrix. The microstructure observation shows that the GNPs-2-6 exhibits a better dispersion in the Cu matrix than GNPs-2-80. The addition of fine GNPs-2-6 lead to ~31% higher tensile strength and approximately same electrical conductivity of the Cu matrix, while the GNPs-2-80/Cu composite only shows a ~15% increase of tensile strength and a lower electrical conductivity than the Cu matrix.

Keywords: Graphene; composites; powder metallurgy; microstructure; properties.

1. Introduction

In recent years, copper matrix composites have received wide attention because of their high strength, good electrical and thermal conductivity. Much effort has been made to explore new reinforcements to further

*Corresponding author.

#To cite this article, please refer to its earlier version published in the *International Journal of Modern Physics B*, Volume 34, 2040045 (2020), DOI: 10.1142/S0217979220400457.

improve the mechanical property while keeping high electrical conductivity of copper matrix composites. Mono- and multi-layer graphene materials are considered as ideal reinforcements in metal matrix composites due to their dazzling properties such as ultrahigh strength, high modulus, excellent electrical and thermal conductivity.[1] The key issues for preparation of high performance graphene reinforced metal composites are to disperse graphene uniformly in the metal and enhance the interfacial bonding strength between graphene and metal. Powder metallurgy method and molecular-level mixing method are reported to be feasible to solve the graphene dispersion and interface problems.[2–4]

Graphene possesses ultrahigh strength along the in-plane direction but low strength in through-thickness direction and is a typical anisotropic material. Therefore, the alignment of graphene may result in a significant improvement of strength and simultaneously maintains the excellent ductility and electrical conductivity of the metal matrix.[5–7] However, upto now, there are few reports on preparation and properties of aligned graphene-reinforced metal composites due to the difficulties to realize a good alignment of graphene in the metal matrix.[5]

In this paper, graphene nanoplatelets (GNPs)/Cu composites were prepared by using two grades of GNPs and flake copper powder as raw materials. GNPs were aligned conveniently in copper matrix due to the self-assembly of flake powder. The effect of GNPs lateral size on the microstructure and properties of GNPs/Cu composites was studied.

2. Experimental

GNPs/Cu composites were prepared using powder metallurgy method. Flake Cu powder and two grades of GNPs (JCNANO Tech Co., Ltd., China) were used as raw materials. Size parameters of Cu and GNPs powders are shown in Table 1. First, Cu and GNPs were dispersed together in ethanol under ultrasonication for 2 h to form a homogeneous suspension. The suspension was filtered and dried. After that, the powder samples

Table 1. Size parameters of raw powders.

Raw materials	Average thickness/nm	Average lateral size/μm	Purity/%
Cu	~100	~10	99
GNPs-2–80	~2	~80	99.7
GNPs-2–6	~2	~6	99.7

were mixed thoroughly by ball milling under a rate of 160 rpm for 5 h with the sample/ball weight ratio of 1:8. Finally, the 10 vol.% GNPs/Cu composite powder was molded into pie shape of Φ40 mm × 5 mm by hot press sintering under vacuum at 650°C for 30 min under a pressure of 35 MPa. For comparison, pure copper specimens were prepared using the same processing parameters.

Scanning electron microscopy (SEM) equipped with energy-dispersive X-ray spectroscopy (EDS) and transmission electron microscopy (TEM) were carried out to observe microstructure of samples. The electrical conductivity was measured and calculated using four-point contact method and the measurement for each sample was repeated nine times. Tensile tests were conducted on a universal testing machine with the crosshead speed of 0.6 mm/min. The test samples with a gauge length of 10 mm and cross-section of 2 mm × 1.5 mm were machined from the sintered billets. Three samples were tested from each sintered billet for tensile tests to minimize the error.

3. Results and Discussion

Figure 1 shows the SEM micrographs of as-received powders. The flake-like and irregular nature of the copper particles can be observed in Fig. 1(a). The typical wrinkle morphology of multilayered GNPs-2–80 is shown in Fig. 1(b). It shows clearly the GNPs include few layers of two-dimensionally arranged ultrathin graphene nanosheets which stacked upon each other. Figure 1(c) gives the stack-like morphology of GNPs-2–6. Compared with GNPs-2-80, GNPs-2–6 exhibit much smaller average lateral size.

Figure 2 shows the SEM micrographs of the GNPs/Cu composite powders prepared by ball milling. GNPs sheets are observed in the mixed powders and the dispersed locations are illustrated by arrows in Fig. 2. As shown in Figs. 2(a) and 2(c), the GNPs-2–80 and GNPs-2–6 are homogeneously mixed with flake Cu particles without obvious agglomeration. An ellipse in Fig. 2(b) indicates cracks in the edge of GNPs-2–80 sheets, which may be resulted from the repeated deformation, cold welding and fracture by the ball milling.[3] The cracks can be the potential weak spot when external force is applied to the composite. In contrast, GNPs-2–6 sheets are compact with no observable cracks. During the late or steady-state stage of ball milling, the work-hardened flake powders will break massively into small particles.[8] The morphologies and sizes of the ball-milled powders are similar to the raw Cu powder, indicating that the

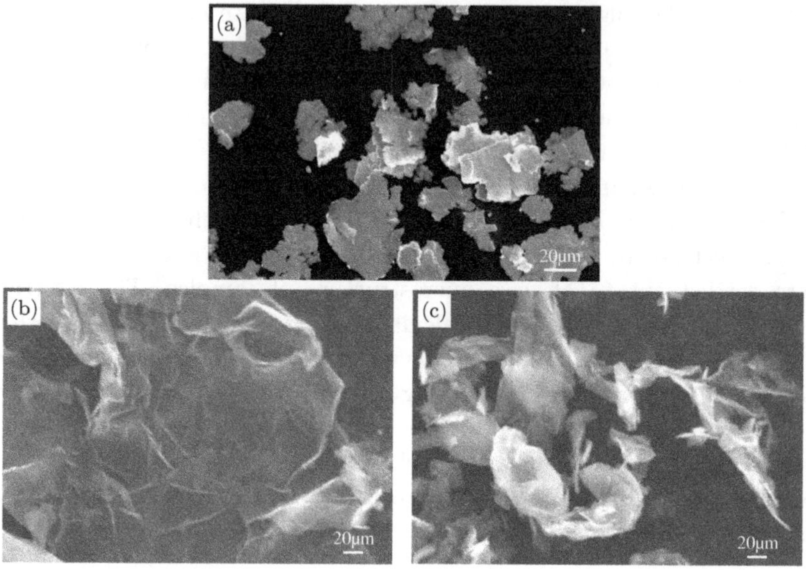

Fig. 1. SEM micrographs of raw GNPs and Cu powders: (a) Cu, (b) GNPs-2–80 and (c) GNPs-2–6.

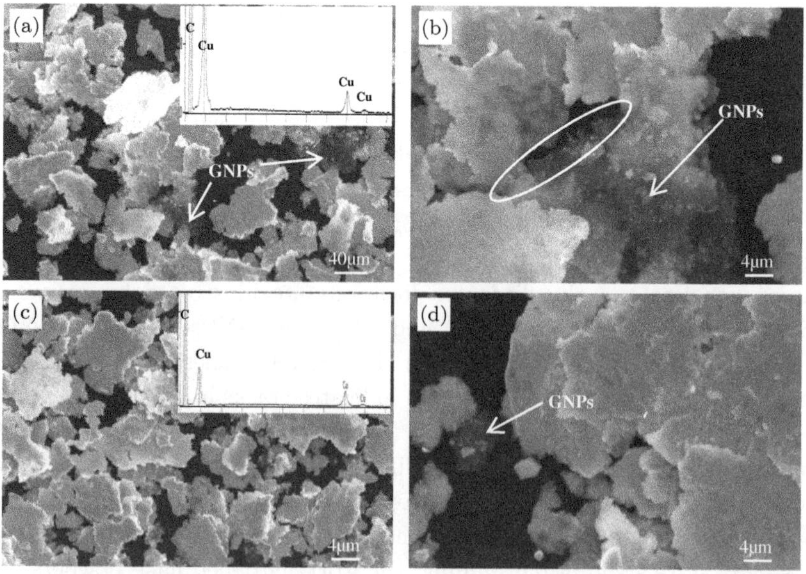

Fig. 2. (Color online) SEM micrographs and EDS of GNPs/Cu composite powders after ball milling: (a, b) GNPs-2–80/Cu powder and (c, d) GNPs-2–6/Cu powder. EDS results were from the marked area in (a, c).

ball milling condition used in this study corresponds to the initial ball milling stage.

EDS analysis in Figs. 2(a) and 2(c) shows the results from the marked area, indicating the presence of C and Cu elements in GNPs/Cu composite powders. Element C is correlated with GNPs. It indicates some Cu particles are covered by GNPs sheets. This coverage can lead to an efficient reinforcing effect of GNPs in the composites.

The microstructure images of the transverse cross section of the sintered Cu and GNPs/Cu composites samples are shown in Figs. 3(a)–3(c). The microstructure in Fig. 3(a) shows that the Cu prepared from flake Cu powder consists of banding grains and many voids, which have a negative influence on the mechanical and electrical properties. In contrast, layered microstructure comprised flat Cu grains formed in the GNPs-2–80/Cu and GNPs-2–6/Cu samples [Figs. 3(b) and 3(c)]. The GNPs sheets are uniformly distributed and well aligned in the Cu matrix. Less voids are observed in GNPs/Cu composites than pure Cu. Thus, the addition of GNPs sheets can promote the sintering process and enhance the density. In addition, the distribution of GNPs-2–6 is much better than GNPs-2–80, which can be ascribed to the difference in lateral size of the grades of GNPs.

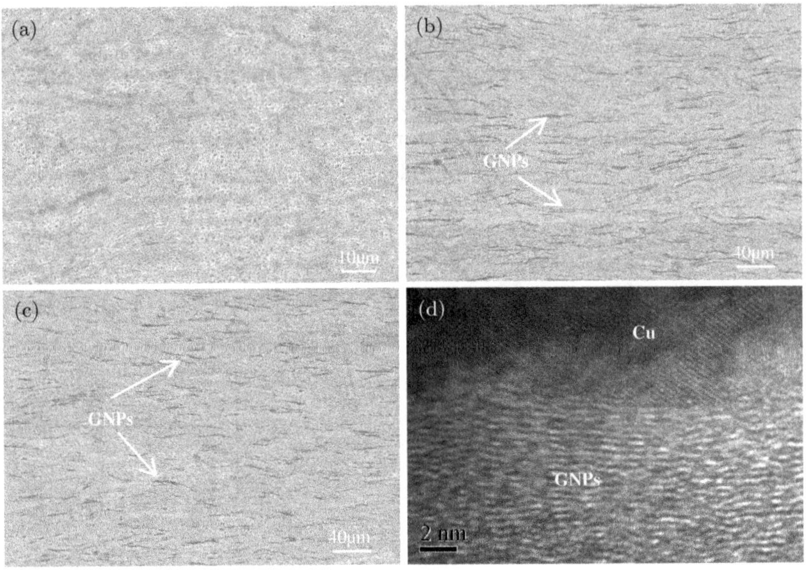

Fig. 3. SEM micrographs of (a) pure Cu, (b) GNPs-2–80/Cu composite, (c) GNPs-2–6/Cu composite and (d) HRTEM image of GNPs-2–6/Cu composite interface.

The interface microstructure of GNPs-2–6/Cu is shown in Fig. 3(d). The HRTEM image shows that no gaps are observed at the interface zone, demonstrating well-contacted interface formed between GNPs and the Cu matrix. The interface transition region can be also observed, indicating the metallurgical bonding at atom scale. The good bonding interface is beneficial to improving the strength of GNPs/Cu composites through strengthening mechanisms such as thermal mismatch, effective transfer load and interaction with dislocations.[9]

Figure 4 shows the tensile strength and electrical conductivity of pure Cu and GNPs/Cu composites. As shown in Fig. 4(a), The GNPs-2–80/Cu composite has a tensile strength of 219 ± 3 MPa, ∼15% higher than that of the Cu matrix. In contrast, the tensile strength of GNPs-2–6/Cu composite is further improved to 249 ± 9 MPa (∼31% enhancement over the Cu matrix). The enhanced strengths indicate that the aligned GNPs sheets are highly effective reinforcement in the Cu matrix. The GNPs-2–6/Al composite shows a better strengthening efficiency than the GNPs-2–80 composite, indicating that graphene lateral size plays a significant role in strengthening the metal matrix.[10] Excessive lateral size can reduce the strengthening effect of graphene.

As shown in Fig. 4(b), the electrical conductivity of the as-prepared samples is relatively low due to the existence of voids (Fig. 3). However, the GNPs-2–6/Cu composite still shows a 2.6% increase of conductivity compared to pure Cu. Our work suggests that graphene can be aligned conveniently in metal matrix by self-assembly of flake powder; and an appropriate lateral size is beneficial for preparing graphene reinforced metal composites with good strength and electrical conductivity.

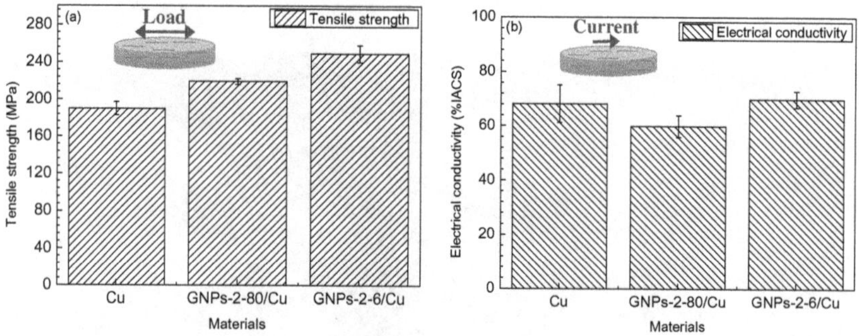

Fig. 4. (Color online) Tensile strength and electrical conductivity of as-prepared pure Cu and GNPs/Cu composites.

4. Conclusions

In summary, copper matrix composites reinforced with two grades of GNPs were prepared by hot pressing. A layered microstructure comprising aligned GNPs was formed in the GNPs/Cu composites by self-assembly of flake powder. The graphene lateral size plays a significant role in strengthening the metal matrix and improving electrical conductivity.

Acknowledgments

The support from the National Natural Science Foundation of China (No. 51801014), the Changzhou Basic Research Program (CJ20179044), the Jiangsu Collaborative Innovation Center for Cultural Creativity (XYN1701) and the Scientific Research Foundation of Changzhou Institute of Technology (YN1715) are gratefully acknowledged.

References

1. C. Lee *et al.*, *Science* **321**, 385 (2008).
2. J. Wang *et al.*, *Scripta Mater.* **66**, 594 (2012).
3. K. Chu *et al.*, *Phys. Status Solidi A* **211**, 184 (2014).
4. J. Hwang *et al.*, *Adv. Mater.* **25**, 6724 (2013).
5. K. Chu *et al.*, *Mater. Sci. Eng. A* **713**, 269 (2018).
6. D. B. Xiong *et al.*, *Sci. Rep.* **6**, 33801 (2016).
7. M. Cao *et al.*, *Carbon* **117**, 65 (2017).
8. W. J. Kim *et al.*, *Carbon* **69**, 55 (2014).
9. R. P. Bustamante *et al.*, *J. Alloy. Compd.* **615**, S578 (2014).
10. M. Zhao *et al.*, *Scripta Mater.* **139**, 44 (2017).

https://doi.org/10.1142/9789819818433_0020

Chapter 20

Solar Energy Harvesting with Ultra-Broadband Metamaterial Absorber[#]

Mehmet Bagmanci[*], Muharrem Karaaslan[*], Emin Unal[*], Oguzhan Akgol[*], Mehmet Bakır[†] and Cumali Sabah[‡,§,¶]

[*] *Iskenderun Technical University, Department of Electrical and Electronics Engineering, Iskenderun, Hatay 31200, Turkey*

[†] *Department of Computer Engineering, Bozok University, Yozgat, Turkey*

[‡] *Middle East Technical University, Northern Cyprus Campus, Department of Electrical and Electronics Engineering, Kalkanli, Guzelyurt, 99738, TRNC/Mersin 10, Turkey*

[§] *Kalkanli Technology Valley (KALTEV), Middle East Technical University, Northern Cyprus Campus (METU-NCC), Kalkanlı, Guzelyurt, TRNC/Mersin 10, Turkey*

[¶] *Sabah@metu.edu.tr*

In this study, a novel metamaterial absorber (MA) is designed and numerically demonstrated for solar energy harvesting. The structure is composed of three layers with different thicknesses. The interactions of three layers bring about the plasmonic resonances. Although the main operation frequency of the structure is chosen between 430 and 770 THz, which is the visible light regime, the proposed structure is also investigated in the ultra-violet (UV) region. One can see from the results that the proposed structure carries nearly perfect absorption capacity that is more than 91% at the whole visible light spectrum. The proposed structure even has the absorption capacity of 99% between 556 and 657 THz. In addition, the designed MA is also investigated in terms of its polarization and angle independency. Results show that the proposed structure is independent from the polarization and incident angles. Lastly, the designed

[¶] Corresponding author.

[#] To cite this article, please refer to its earlier version published in the *International Journal of Modern Physics B*, Volume 33, 1950056 (2019), DOI: 10.1142/S0217979219500565.

structure can be considered to be used in solar energy applications as a harvester since it has an ultra-broadband absorption characteristic in visible light regime.

Keywords: Metamaterials; absorber; energy harvesting.

1. Introduction

Renewable energy sources, especially the solar energy, can be used for solving energy related environmental problems as well as reducing the global dependence on fossil fuels. This energy source is flexible, environmentally friendly, cost effective and widespread. It can be used in heating–cooling, electricity generation, etc.[1,2]

Since solar energy is free, nonpolluting, has no moving parts that could break down and only requires a little maintenance and no supervision, researchers' interest in solar energy has been increasing rapidly. Major part of the researches have been oriented on the design and manufacturing of solar cells, improving the efficiency of solar cells and the ability of light absorption.

It is possible to absorb and convert solar radiation into electrical energy by using solar cells. This conversion is possible by photovoltaic effect that occurs under the solar radiation. Solar radiation causes electrons from the atoms in the semiconductor material to be loose and creates electron–hole pairs. The electrons are then captured in the form of electric current.[3,4] The wavelength of the solar radiation is placed between 300 to 3000 nm, in other words, it is between 100 to 1000 THz.[5,6]

The early generation solar cells which are made out of silicon wafer solar cells that have large areas are still in the market. Since processing silica (SiO_2) to produce pure silicon is costly, a second generation technology based on thin films emerged. By this technology, it is expected to produce 100 μm thick wafers in the future.[7] The production of solar cells with small thickness leads to a decrease in the absorption of light. For this reason, light trapping techniques were developed.[8,9]

Plasmonics or plasmonic metamaterials are another type of artificial material. They have exotic properties such as negative refractive index.[10,11] Recently, solar cell-based metamaterials have been developed to be used in a wide range of frequencies to absorb solar radiation.[12–24]

These conducted studies investigated the tunable perfect absorber with graphene-hexagonal gold cylinder array on a ground plate.[12] Seventeen recent articles on EM absorbers based on metamaterials in different frequency bands have been reviewed in Refs. 13. In Ref. 14, selective thermal emitters based on metamaterial perfect absorbers were demonstrated for the first time. They experimentally realized a narrow band mid-infrared (MIR) thermal emitter. In Ref. 15, the method of EM reduction was presented and the effects of position, distance, and size of metamaterials were analyzed. A new triangular metamaterial achieved a 1.0923 W/kg for SAR 1 gm which is 45.44% reduction of the initial SAR value for the case of 1 gm SAR. Dual resonant broadband terahertz metamaterial based on a modified split ring resonator was designed, manufactured and tested in Ref. 16. The designed sample is fabricated on a gallium arsenide layer, and the experimental studies were performed in Terahertz Time-Domain Spectroscopy.

Wang *et al.* investigated a perfect selective metamaterial solar absorber designed by using two-dimensional tungsten gratings on a thin dielectric.[17] Another metamaterial absorber-based structure having a two-dimensional pyramidal shape at infrared waveband for normal incidence was improved by Liang *et al.*[18] Cao *et al.* studied a polarization insensitive broadband perfect metamaterial absorber composed of an array of thin Au squares at visible frequency regime.[19] A titanium nitride (TiN) plasmonic material-based polarization angle independent broadband metamaterial absorber structure for solar thermo photovoltaic application was designed by Wang *et al.*[20] Similarly, Azad *et al.* proposed a broadband, polarization independent wide angle solar absorber based on a metallic metasurface architecture in the visible and near-infrared range.[21] Abdel-Latef *et al.* realized the funnel-shaped NW consisting of a cylinder over a conical unit. They aimed to maximize the optical absorption over a large wavelength range and hence the electrical efficiency by increasing the number of resonance wavelengths or by enlarging the resonance wavelength range.[22] Yu *et al.* reviewed recent works on SiNWs. Following that, various techniques aiming to achieve high quality nanowires at low costs are introduced.[23] Lei *et al.* proposed periodic array structures for visible region application by using titanium-silica (Ti-SiO_2).[24] Litchinitser reviewed a subset of recent progress in theoretical and experimental studies of nonlinear optical properties and phenomena enabled by the unique properties of engineered optical media and new regimes of nonlinear optical interactions enabled by judiciously designed metamaterials.[25] In contrast to the previous metasurface carpet cloak in single band, a macroscopic metasurface carpet cloak was proposed to control the reflection phases at several independent frequencies simultaneously in Ref. 26. Ma *et al.* experimentally demonstrated the concept of hyper structured illumination. A ~80 nm resolution has been achieved in a well-known Ag/SiO_2 multilayer hyperbolic metamaterial system by using a low numerical aperture objective (NA = 0.5), representing a 6-fold resolution enhancement of the diffraction limits.[27] Dinçer *et al.* proposed a metamaterial absorber based on a cylindrical waveguide with polarization independent sensor applications with 99.7% absorption.[28] Sabah *et al.* developed a chiral metamaterial-based temperature sensor to show biosensor applications of metamaterials.[29] Akgol *et al.* developed a metasurface to realize a polarization converter to transform linearly polarized signal into circularly polarized one by using numerical and experimental techniques.[30] Altıntaş *et al.* realized a polarization converter that has a bandwidth of 800 MHz axial ratio.[31] Dincer *et al.* realized a polarization independent metamaterial absorber that has an absorption rate of 99.99% at 5.48 GHz and 99.92% at 0.865 THz.[32] Bakır *et al.* realized an absorber that operates at industrial, scientific and medical bands with sensor applications.[33] Karaaslan *et al.* designed, characterized and fabricated an electromagnetic bandgap (EBG) structure composed of three different scale planar rectangular mushrooms for low profile antenna applications.[34] Ding *et al.* reviewed a broad subclass of metasurfaces, gradient metasurfaces, which are devised to exhibit spatially varying optical responses resulting in spatially varying amplitudes, phases and polarizations of scattered fields.[35] In addition, properties of complex silica embedded nickel nanoparticles are given by Mahoney *et al.*[36]

Metamaterial absorber can be designed to operate in infrared region as in Refs. 37–41. Perfect metamaterial absorbers can be employed for solar cell applications. Rufangura and Sabah designed a metamaterial that gives a nearly perfect absorption (99.94%) with a bandwidth of 23.4% in visible spectrum.[42] Deshmukh *et al.* demonstrated long-range (~160 nm) direct energy transfer between donor quantum dots and acceptor dye molecules through the use of an optical topological transition (OTT) in a metamaterial.[43] Abdelraouf *et al.* realized a solar cell application by using periodic boundary condition and the effect of replacing molybdenum with refractory plasmonics titanium nitride (TiN). They showed that after using TiN, enhancements in light absorption are seen.[44] Sharma *et al.* performed a solar cell study by using dye sensitized solar cells and use of nanomaterials.[45] Zhang *et al.* enhanced optical absorption in a thin organic layer by embedding into magnetic metamaterials. Light absorption has been obtained in a broadband range from 350 to 750 nm in solar spectrum.[46]

In this study, we design, characterize and analyze a new kind of metamaterial absorber operating in visible spectrum to utilize solar energy effectively. The proposed structure consists of an array of nickel–SiO_2–nickel combinations and it offers polarization and angle independency, broadband and near-perfect absorption characteristics for visible region applications. Although the proposed structure has a complex geometry when compared with some studies (especially, Refs. 37–41), it has a special purpose with regard to the material type used. Nickel (Ni) is a predominant material in common high-temperature alloy metals. Thus, this study is aimed to be used in especially high-temperature applications of solar energy absorber. Besides, the reason why we choose the Ni and SiO_2 combination is to obtain cost-effective, durable and easy producible metamaterial absorber structure.[38] Although the proposed solar cell has a complicated structure and overall thin thickness brings great difficulties to the actual processing, current state of the art may overcome these drawbacks as reviewed by Yu *et al.*[47] High density nanocomposite can lead to a broadband absorption due to excitation of localized plasmon resonances of the nanoparticles by visible light.[48] Gold, copper and silver nanocomposites can be employed for broadband perfect metamaterial absorbers as realized by Hedeyati *et al.* and Bisvas *et al.*[48–50] Elhabri and co-workers applied Au/SiO_2 nanocomposite instead of metallic structure in perfect metamaterial absorber which results in 100% absorption over a broadband range from Ultra Violet (UV) to near-infrared region.[50] Proposed structure employs nickel similar to gold, copper and silver. By using Elhabri and co-worker's technique, it can be applied to SiO_2. They have used a cylindrical vacuum chamber for sputtering of the metal and polymeric film. By installing two magnetrons with an angle of 50°, sputtering was done in order to make a nanocomposite out of a copper and PTFE targets. In order to acquire a uniform thickness for the film and homogeneous metal distribution for the composite, all depositions were done on the samples attached to a rotatable sample holder.[50]

The Ni is also compared several metals to observe absorption ability of the proposed structure. The new design shows a perfect absorption between 550 (545 nm) and 650 (461 nm) THz which is validated numerically. Electric field and surface current distributions are analyzed numerically to explain physical mechanism of the proposed design. Moreover, the dimensions of the unit cell and layer thickness are optimized by using the genetic algorithm method. Rest of the paper is organized as follows: design and numerical setup are explained in Sec. 2; simulation results are presented in Sec. 3; and finally, the paper is concluded in Sec. 4.

2. Design and Numerical Setup

The design of the proposed structure is shown in Fig. 1. Size of the structure and the thicknesses of metals and dielectric layer are given in Table 1. Resonator composed of squares which are aligned on diagonal, horizontal and vertical lines of the structure is demonstrated in Fig. 1(a). Metallic parts of the structure are Nickel (Ni) and the dielectric part of the structure is silicon dioxide (SiO_2) as shown in Fig. 1(b). The reason for choosing nickel is its better absorption efficiency compared to the other metals at visible region. Dielectric parameters which are complex and dependent on the frequency for nickel and silicon dioxide are needed for the simulation studies which are obtained from the library of the CST Microwave Studio for the specific visible light frequency range. The boundary conditions of the structure are assigned as perfect electric conductor (PEC) and perfect magnetic conductor (PMC) in order to achieve periodicity for the transverse electromagnetic (TEM) incident case. The PEC and PMC boundaries are perpendicular to the electric and magnetic field components of the electromagnetic wave, respectively. These boundary conditions are only valid for the TEM wave to exhibit periodicity for unit cell structure. The solar energy is applied from the open add space wave port with a separation distance to prevent near-field effect between EM wave and the absorber. The backside of the solar absorber is chosen to be metal to block signal transmission to the backside of the cell. The solar absorber is designed to be symmetric in order to provide exactly the same response for

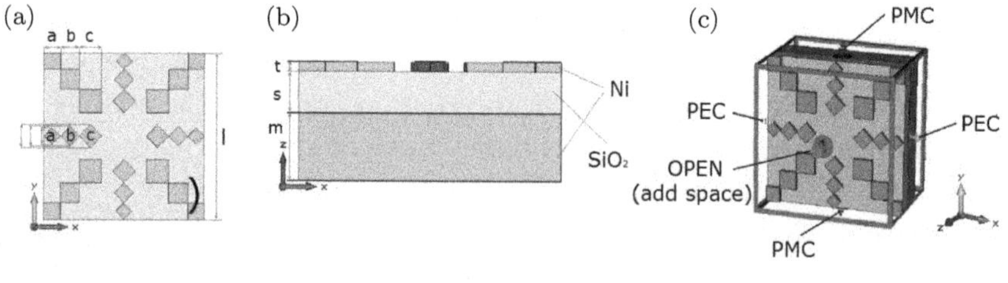

Fig. 1. (Color online) Unit cell geometry for the designed structure. (a) Front view, (b) side view and (c) boundary conditions.

Table 1. Typical sizes for the designed structure.

Parameter	a	b	c	l	t	s	m
Value (nm)	50	60	70	500	15	60	100

solar energy with different incident angles and polarizations. The mentioned dimensions are determined by a parametric study and by the genetic algorithm approach tool of CST to restrain all the incident solar energy in the structure for all visible light wavelengths.

One of the reasons of choosing these geometries is the tuning option; as it can be seen from Fig. 1, there are two different square shapes with three elements inside placed on eight different places on the substrate. The placement of these unit cells has been decided by trying and evaluating S parameters given by the simulation program. The other reason related with the geometry depends on the genetic algorithm which is used in the simulation program. When we entered the substantial objectives which are related with the absorption frequency and its level and band, the algorithm evaluated the dimensions after long calculations and fine tuning of dimensions on these geometries finished.

Production of Nickel–SiO_2–Nickel combination with thicknesses mentioned in this paper is possible by using complex engineered nanomaterials and metasurface as in Refs. 35 and 36. These materials are a rapidly developing class of structurally and compositionally complex materials that is expected to generate the next generation of nanomaterials.

3. Simulation and Discussion

In this study, simulations are carried out first for identifying the unit cell and for obtaining a better understanding in the absorption mechanism of the investigated structure. The effects of the inner, middle and outer diagonal square resonators are demonstrated one by one for the determination of the operation frequency. Effects of the horizontal and vertical square resonators are also investigated at the same frequency interval. Finite integration technique-based CST Microwave Studio, commercial electromagnetic simulation software, was used to explore the absorption behaviors of the designed structure. Boundary conditions are set in a way that the electric boundary is in the x direction and the magnetic boundary is in the y direction while open boundaries should be selected in the z direction. Besides, it is assumed that the incident electromagnetic wave has electric and magnetic components which are polarized in x and y directions, respectively. The propagation has the direction along the positive z axis. Since the visible frequency range (430–770 THz) or wavelength (697–389 nm) has high solar radiation, this frequency range is chosen for the study. In addition to the visible frequency region, infrared (100–430 THz) and UV (770–1000 THz) frequency ranges are also investigated for the proposed structure.

Absorption behavior of the designed structure which will be shown as $A(w)$ can be calculated with the help of the reflection and transmission parameters as

$$A(w) = 1 - R(w) - T(w).$$

In order to achieve the highest absorption level, reflection $R(w) = |S_{11}^2|$ as well as the transmission $T(w) = |S_{21}^2|$ values should be minimized. Since $T(w)$ is completely blocked by Nickel plate located at the back side of the cell in Fig. 1(b), the transmission of the proposed structure is negligible. Therefore, the absorption of the designed structure only depends on the reflection values $R(w)$ which is the only term left after assuming $T(w)$ as negligible. Particularly, the optimization and simulation studies are analyzed by considering the new updated version of the equation for the absorption response of the proposed design which is $A(w) = 1 - R(w)$, where $|S_{21}^2| = 0$ thus, $T(w) \to 0$.

Figure 2 shows both the absorption and reflection characteristics of the designed metamaterial solar absorber. Unlike the other studies in the literature about broadband MA, the proposed structure can be seen to have a wide bandwidth and best absorption feature in the whole band of visible regime. Namely, the structure has got the absorption value of 99.71% at 608 THz (493 nm) and an absorption level above 99% in the frequency range of 556 (539 nm) and 657 THz (456 nm). Furthermore, the proposed structure has a high absorption capability which is above 91%

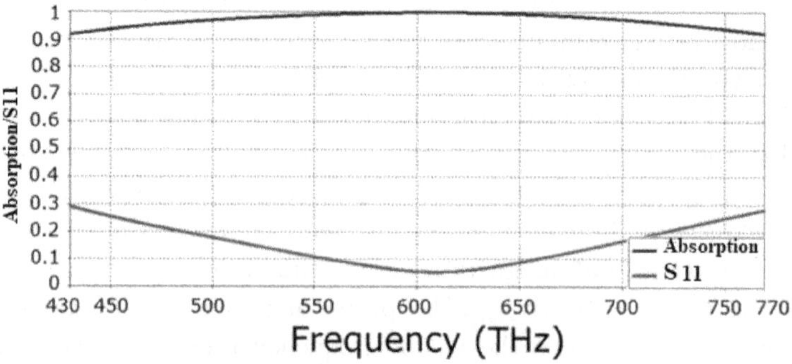

Fig. 2. (Color online) Absorption and reflection behaviors of the designed structure.

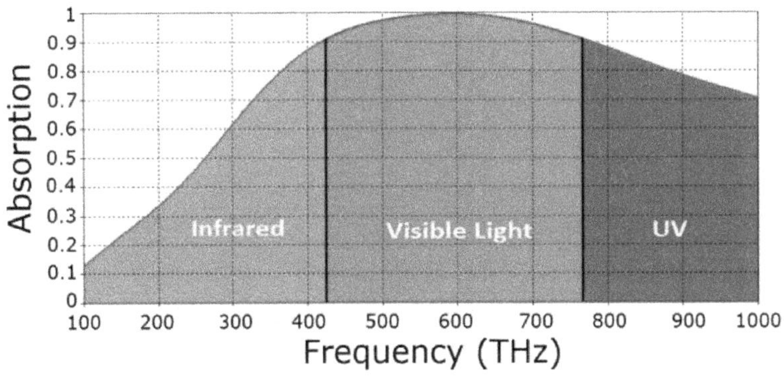

Fig. 3. (Color online) Absorption characteristic of the proposed structure at all sun light regime.

at all frequencies in the visible light region. It means that the structure has the capacity to confine all the solar energy between Ni–SiO_2 intervals. Hence, the total incident energy is allowed to penetrate into the structure and has potential to convert electrical current.

In addition to the visible regime, the absorption behavior of the designed structure is also examined at infrared and UV frequency regions which are demonstrated in Fig. 3. For the infrared region, the absorption percentage is above 90% between 424 (707 nm) and 430 THz (697 nm). This is a very good performance for an infrared detector operating in the infrared region of the spectrum. Infrared detectors have wide applications in area such as infrared motion detectors, infrared petroleum exploration, gas leak detection, infrared anaesthesiology testing, wildlife observation and many more from military applications to medical and automotive fields and the fabrication of an IR detector with a good sensitivity needs an efficient IR absorber in the mentioned frequency band. For the UV region, the absorption percentage of the proposed structure is above 90% between only 770 (389 nm) and 774 THz (387 nm). Additionally, the proposed MA has 70% absorption which is the minimum magnitude of the absorption characteristic at 1000 THz (299 nm). Therefore, we can infer that the absorption level of the designed structure is high only in the invisible section of the spectrum, but it has high absorption efficiency in a narrow band in the infrared and UV regions. However, different structures with various dimensions can be designed to realize high absorption ratios for either infrared or UV regions for future studies.

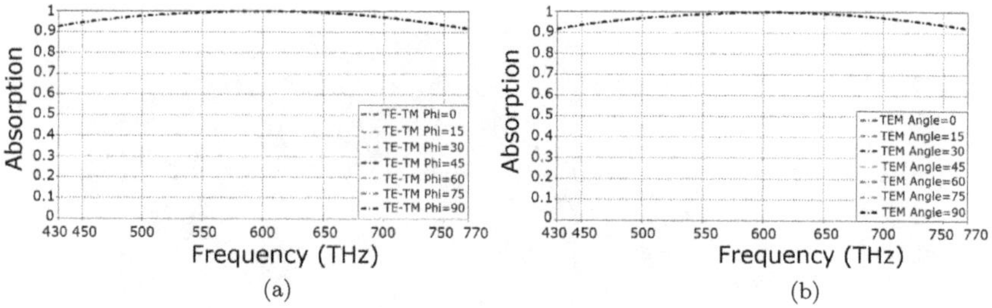

Fig. 4. (a) The absorptivity responses for different incident angles under both TE and TM polarized waves and (b) TEM mode.

Proposed structure is also independent from the phi angle for TE and TM polarization which is shown in Fig. 4(a) and it shows an angle independent characteristic for TEM mode which is shown in Fig. 4(b). This feature of the structure makes it unique in point of angular independency from the other structures in the literature since it is not sensitive for the TE and TM polarization as well as TEM mode. Therefore, the proposed structure can be used as a metamaterial-based solar absorber in the entire visible light region with an angular and polarization independency. This provides flexibility for scientists investigating solar absorber.

Second, the absorption efficiency of the proposed structure for different metals is compared with nickel. In Fig. 5, the comparison of nickel with aluminium, copper, silver and gold is shown. It is clearly seen that the absorptivity of the structure with nickel is better than others with an absorption value higher than 90% in the entire range of visible spectrum. The reason is that the imaginary part of dielectric constant of nickel is smooth and higher than other metals in this frequency range. The worst absorption value is obtained for silver of which the imaginary part is very low due to its better conductivity and low losses. Structure of the proposed design also shows better results than other structures in the literature, where aluminium, copper, gold, nickel and silver are used as metals thanks to their perfect symmetrical geometry. The nickel-based metamaterial solar absorber can be used in the entire frequency range of visible light spectrum with a leading performance compared to the other metals.

Besides this, it can be concluded that exactly the entire part of the incident energy is absorbed by metals. Since the imaginary part of permittivity of silicon dioxide is around 2×10^{-5} (Ref. 51) and it is much less than that of even silver,[52] the effect of silicon dioxide to the absorption

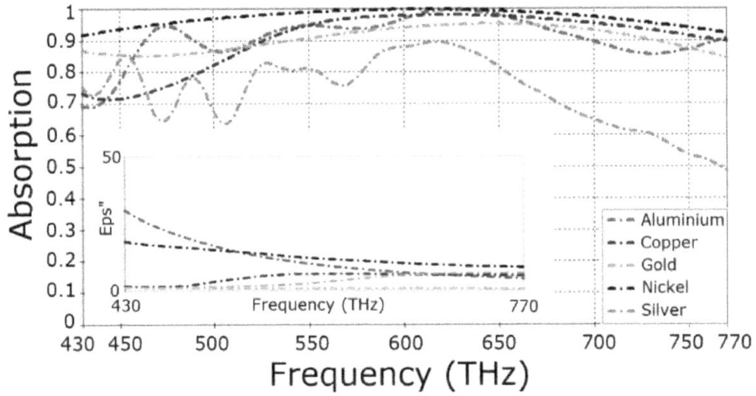

Fig. 5. (Color online) Comparison of absorption percentage for different metal usage.

characteristics of overall system can be neglected. Hence, almost the entire penetrated power is absorbed by the metallic parts in the system. Although Ni is a poor plasmonic material and absorbs most part of the energy in visible range, silver-based approach is used to observe obviously the difference of absorption changes because silver has low loss characteristic with respect to the other metals used in this comparison shown in Fig. 5.

In this part, the resonators of the proposed structure have been divided into two groups to specify these elements. The first group consists of the square patches at the diagonal line of the structure and the second group consists of the square patches at the vertical and horizontal lines of the structure. First, the dependence of absorption characteristic of the first group is examined. The simulations are realized for the TEM incident case with the mentioned boundary conditions. The dimensions of each individual part are the same with the overall structure dimensions shown in Table 1. The individual and the combined response of the diagonal square patches are illustrated in Fig. 6. The outer square resonators can be seen to have minimum absorption percentage and the combination of squares has maximum absorption percentage. The maximum absorption magnitudes of only outer, middle and inner squares are 81.54% at 685.6 THz (437 nm), 86.68% at 644.8 THz (465 nm) and 89.61% at 632.8 THz (473 nm), respectively. The maximum absorption percentage of the combination of all squares is 96.28% at 622.8 THz (481 nm). We can infer that while outer squares have maximum magnitude at high frequencies and have a low absorptivity, inner squares have maximum magnitude at low frequencies

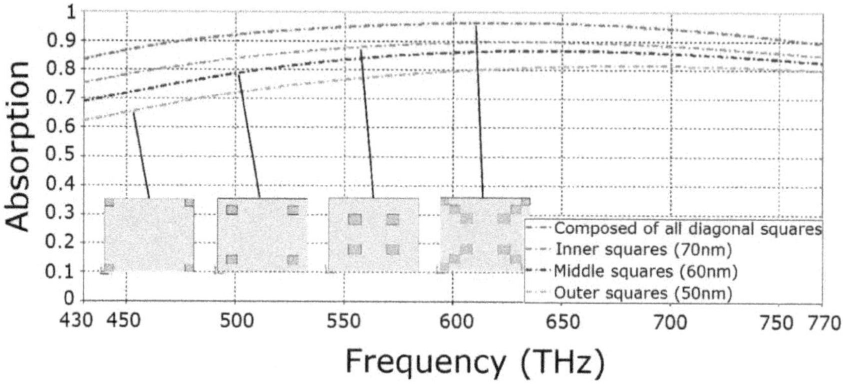

Fig. 6. (Color online) The individual and combined responses of diagonal square patches.

and have a high absorptivity. The combination of diagonal squares in the same way has the highest absorptivity but it has maximum magnitude at the lowest frequency which is compared to the individual responses. In this respect, the combination of all diagonal squares can be used as a solar absorber with a good performance of around 90% absorptivity in the whole range of visible light. This is resulted from matched intrinsic impedance of the proposed structure with the medium hosting the incident wave and the metal shield placed at the backside. Besides, all proposed fractions have both angular and polarization insensitivity due to the central symmetry.

Second, the dependence of absorption characteristic of the first group (diagonal square patches), second group (vertical and horizontal square patches) and the proposed structure are compared. All the dimensions are the same with the lengths given in Table 1. The individual responses of mentioned groups are given in Fig. 7. Results show that the first group patches have 96.28% peak absorption at 622.8 THz (481 nm), the second group patches have maximum absorption of 88.71% at 651.6 THz (460 nm) and the proposed structure has 99.71% maximum absorption at 608.8 THz (492 nm). When all of the resonators are combined, the absorption strength increases due to coupling and plasmonic effect of the resonances. Since the incident wave has TEM characteristic, diagonal patches indicate a better absorption response due to the higher metallic area for E field resonance and higher current flow area for H field resonance. Both first and third structures are good candidates for being used as absorbers in solar absorption applications and in the whole spectrum band of visible light.

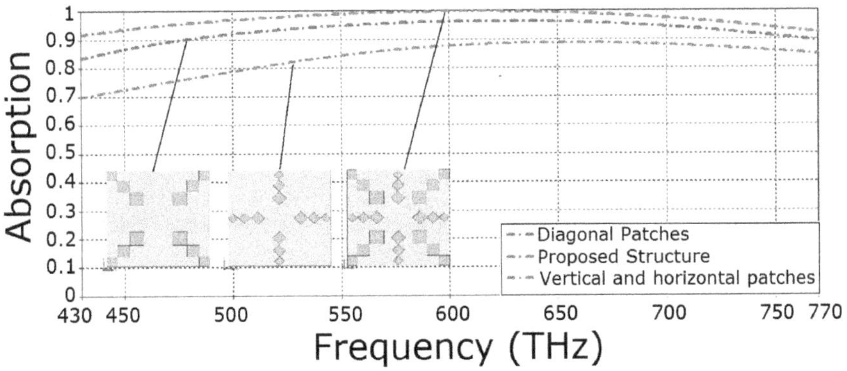

Fig. 7. (Color online) The absorption characteristic of diagonal, vertical and horizontal and proposed structure.

A high percentage of Ni atoms accumulate at the surface which exhibits the special features of Ni metallic nanoclusters like strong absorption band and localized surface plasmon resonance. These features are caused by the quantum confinement effects of Ni metallic nanoclusters similar to the metals used in Refs. 53 and 54.

In the next step, dependence of absorption values of the structure to the geometric parameters is investigated. While other parameters are kept fixed, thicknesses of the resonators and dielectric layers are changed for characterization and better understanding of the operation mechanism of the absorber. All mentioned parameters are altered and simulated synchronously so as to obtain optimum parameters for the proposed structure. Results are demonstrated in Figs. 8(a) and 8(b). It should be noted that black curves in both Figs. 8(a) and 8(b) show the desired results with respect to the absorption percentage. Therefore, these parameters can be selected as optimal parameters for the proposed structure for a perfect absorption.

First, the variation of resonator thickness (m) is investigated which is shown in Fig. 8(a). Except the thickness of 10 nm where the maximum absorption is 97%, the other values of thickness have 99% absorption peak. Peak value reduction is due to lower path dimensions of the wave in the proposed structure. Since the intrinsic impedance of the structure is almost same with the one in free space, the entire incident wave penetrates into the solar absorber. However, reduction of the track increases the leakage of the wave for every return of the wave to the open space face of the structure.

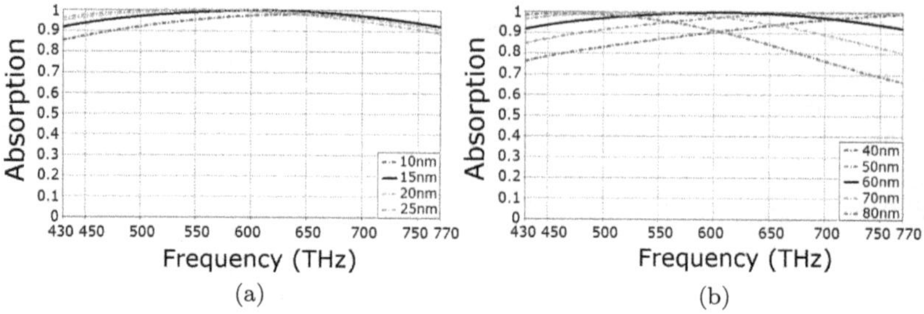

Fig. 8. (Color online) Thickness changes of (a) resonator and (b) dielectric layer.

Whereas, values of 15, 20 and 25 nm thicknesses have 99% absorption at the resonant frequency, minimum absorption values of these thicknesses are 91%, 90% and 88%, respectively. Namely, three thickness values have 99% absorption peak, but minimum absorption value of the desirable one which is black and solid curve has the highest magnitude. In addition, when the resonator thicknesses increase, resonant frequency shifts towards to the low frequencies. This is caused by the reduction of capacitance of the resonator.

Second, the thickness of dielectric layer is examined. The effect of the dielectric thickness to the absorption rate is observed by changing the parameter "s". The response of dielectric layer changes is demonstrated in Fig. 8(b). The dielectric thickness primarily helps the coupling between the electric resonance on metallic resonators and bottom metallic layer. Therefore, this coupling generates a magnetic response and makes the effective impedance of the absorber to be equal to the impedance of free-space (impedance matching). That is why the optimum thickness of the dielectric layer is found as 60 nm abiding by impedance matching. Selection of thickness of dielectric layer is also very important in order to have perfect absorption which can be seen in Fig. 8(b).

Furthermore, in order to understand resonance behavior and the operation mechanism of the metamaterial absorber, electric field, magnetic field and surface current distributions at the minimum operation frequency (430 THz), at the maximum operation frequency (770 THz) and at the maximum absorption frequency point (608.8 THz) are presented in Fig. 9. Magnitude of electric field distributions ($|E|$), x-component of the electric field (E_x), y-component of the magnetic field (H_y), and surface current distribution of the proposed structure at the mentioned frequencies are demonstrated in Figs. 9(a), 9(b), 9(c) and 9(d), respectively. At 430 (697 nm), 608.8 (492 nm) and 770 THz (389 nm), the electric field is mostly

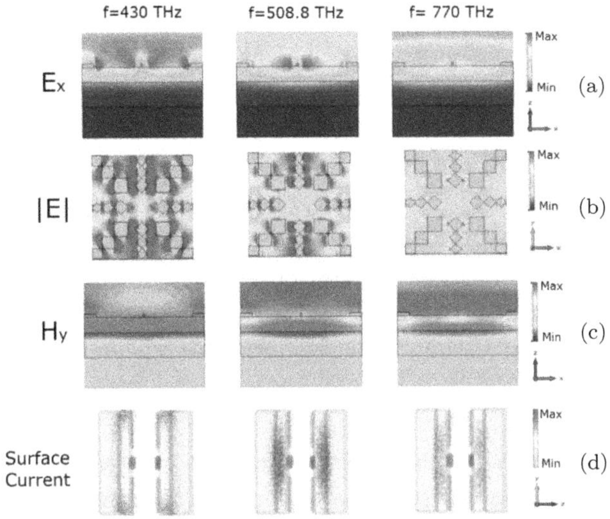

Fig. 9. (Color online) (a) x-Component of the electric field distributions, (b) magnitude of electric field distributions, (c) y-component of magnetic field distributions and (d) surface current distribution of designed structure.

concentrated near the edges of the vertical square, the intersection of the edges of the diagonal square and intersection of the edges of the horizontal square, which are shown Figs. 9(a) and 9(b). However, while the electric field is concentrated at the edges of vertical squares which are in the x direction, it is not concentrated at the edges of vertical squares which are in the y direction. In the same way, while the electric field is concentrated at the edges of horizontal squares which are along the x direction, it is not intensified at the edges of horizontal squares which are in the y direction. As for diagonal squares, the electric field occurred at the intersection of the edges of the diagonal square at the mentioned frequency. Additionally, when the frequency increases, this concentration of electric field lessens, which is clearly seen in Figs. 9(a) and 9(b). Magnetic field distributions of the proposed structure at the operation frequencies are shown in Fig. 9(c). We can see that the coupling between all square resonators and the ground plane crops up with the generation of the magnetic polarity which occurs in anti-parallel distributions. While magnetic polarity is caused by all squares at lower frequencies, magnetic polarity is mainly created by the vertical square at higher frequencies. Lastly, surface current distributions are presented in Fig. 9(d). It can be seen that surface currents occur at both

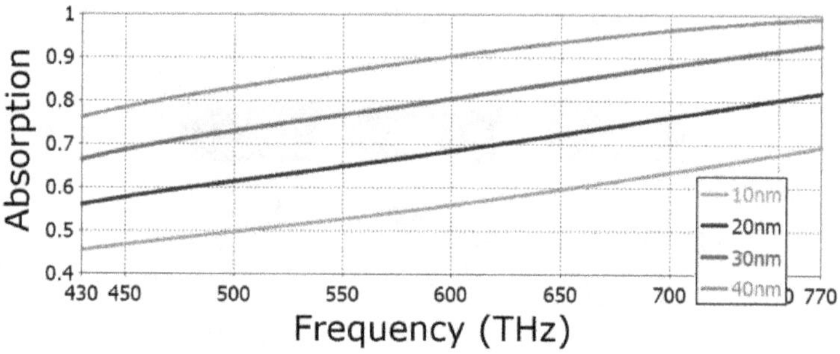

Fig. 10. (Color online) Changes of the absorption value with respect to the dielectric thickness (SiO_2) in the range of visible light.

sides of the structure. While surface currents mostly occurr near the edges of the structure at 430 THz, they are formed at the centre of the proposed structure at 608.8 and 770 THz. However, the surface current density at the frequency of 608.8 THz is more than the one at the 770 THz.

Besides this, the dielectric thickness effect on absorption value is also investigated as shown in Fig. 10. The increment of the dielectric thickness results in an improvement in the absorption potential of the MA structure due to a longer and lossy path between the front and backside of the structure. The absorption level is almost higher than 80% in the whole range of visible light. Therefore, the structure with a thickness of 40 nm can be used to absorb most of the incident energy for the applications of visible light cloaking, stealth technology and solar absorber with proper energy conversion. Another significant point of the structure from this result is that the proposed MA has a potential optical sensor with linear absorption value with respect to the incident wave frequency. Hence, the structure with a dielectric thickness of 20 nm can be used as an optical frequency sensor and color detector which can detect entire optical spectrum.

The proposed study is compared with the other previous works which are achieved by Yang *et al.*,[37] Ma *et al.*,[38] Ling *et al.*[39,40] and Ding *et al.*[41] The comparison is tabulated in Table 2 due to the absorption ratio above 90%. It should be noted that the proposed structure has an extremely wide absorption bandwidth at infrared and invisible regime. Furthermore, it is clearly seen in Table 2 that the structure has the thinnest thickness compared with the other structures. Hence, it can be concluded that the proposed absorber is compatible for solar energy harvesting applications with respect to the operating frequency range and structure thickness.

Table 2. Comparison between the earlier studies in the literature and proposed structure with respect to absorption bandwidth and structure thickness.

Structures	Operation frequency (THz)	Bandwidth (GHz) (absorption > 90%)	Thickness (nm)
Yang *et al.*[37]	62–64 and 69–72	5	200
Ma *et al.*[38]	75–85	10	390
Ling *et al.*[39]	40–70	30	1300
Ling *et al.*[40]	60–115	55	700
Ding *et al.*[41]	175–300	125	260
Proposed structure	430–770	340	175

4. Conclusion

In conclusion, an ultra-broadband MA is designed in this study by a combination of three layers. Unlike the other structures based on multiple layers which are presented in literature, the proposed structure has just three layers, ultra-wide bandwidth and nearly perfect absorption for the whole range of visible light. Besides, the proposed structure is also investigated at infrared and UV frequency regimes. The qualification of the broadband resonances is performed by identifying the responses of each square resonator patch and discovering the origins by using the electric and magnetic field distributions at the minimum operation frequency, maximum operation frequency and the resonant frequency. The investigation of dependency of the geometric parameters which are resonator thickness and dielectric layer thickness on the metamaterial absorber is also presented. As a conclusion, the proposed structure is a polarization and angular insensitive ultra-broadband metamaterial absorber and it can be used in solar energy harvesting, absorbing solar radiations, imaging in the solar spectrum and in the sensing applications.

The proposed structure is designed for future solar absorbers to design solar cells. The absorbed energy by dielectric layer is over 70% in nearly all frequency ranges for all the different metal structures as mentioned above. Hence, this confined energy is ready for the conversion of electrical energy, since the penetrated energy in the dielectric layer goes and returns between the front and backside metallic layer. This conversion can be achieved by embedding appropriate semiconductors in the sandwiched layer. The highly broadband absorber reported here for Ni–SiO_2 can be used in important applications, such as harvesting energy from light, biomedical imaging and sensing.

References

1. N. Muhammad *et al.*, *Materials* **11**, 2315 (2018).
2. S. Chhawchharia *et al.*, *Renewable Sustainable Energy Rev.* **91**, 888 (2018).
3. W. Li and J. Valentine, *Nano Lett.* **14**, 3510 (2014).
4. F. Dincer *et al.*, *Progress Electromagnetic. Res.* **144**, 93 (2014).
5. P. Rufangura and C. Sabah, *IET Optoelectron.* **10**, 211 (2016).
6. M. A. Memon *et al.*, *Renewable Sustainable Energy Rev.* **82**, 2235 (2018).
7. F. Enrichi, A. Quandt and G. C. Righini, *Renewable Sustainable Energy Rev.* **82**, 2433 (2018).
8. J. E. Haverkort, *Appl. Phys. Rev.* **5**, 031106 (2018), https://doi.org/10.1063/1.5028049.
9. N. Zin *et al.*, *Solar Energy Mater. Solar Cells* **183** (2018).
10. M. I. Aslam and Ö. G. Durdu, *Phys. Rev. B* **84**, 195465 (2011).
11. J. Hao *et al.*, *Appl. Phys. Lett.* **96**, 251104 (2010).
12. P. Zhou *et al.*, *Opt. Mater.* **81**, 59 (2018).
13. Y. Cui *et al.*, *Laser Photonic. Rev.* **8**, 495 (2014), doi:10.1002/lpor.201400026.
14. X. Liu *et al.*, *Phys. Rev. Lett.* **107**, 045901 (2011).
15. M. R. I. Faruque, M. T. Islam and N. Misran, *Progress Electromagnetic. Res.* **124**, 119 (2012).
16. W. Guo *et al.*, *Progress Electromagnetic. Res.* **134**, 289 (2013).
17. H. Wang and L. Wang, *Opt. Exp.* **21**, A1078 (2013).
18. Q. Liang *et al.*, *Adv. Opt. Mater.* **1**, 43 (2013).
19. T. Cao *et al.*, *Sci. Rep.* **4**, 3955 (2014).
20. H. Wang *et al.*, *Photonic. Res.* **3**, 329 (2015).
21. A. K. Azad *et al.*, *Sci. Rep.* **6**, 20347 (2016).
22. G. Y. Abdel-Latif *et al.*, *J. Photonic. Energy* **7**, 047501 (2017), https://doi.org/10.1117/1.JPE.7.047501.
23. P. Yu *et al.*, *Nano Today* **11**, 704 (2016).
24. L. Lei *et al.*, *Opt. Exp.* **26**, 5686 (2018).
25. N. M. Litchinitser, *Adv. Phys. X* **3**, 1 (2018), doi:10.1080/23746149.2017.1367628.
26. C. Wang *et al.*, *Opt. Exp.* **26**, 14123 (2018).
27. Q. Ma *et al.*, *ACS Nano* **12**, 11316 (2018), doi:10.1021/acsnano.8b06026.
28. F. Dincer *et al.*, *Mod. Phys. Lett. B* **30**, 1650095 (2016).
29. C. Sabah *et al.*, *J. Electromagnetic Waves Appl.* **29**, 2393 (2015).
30. O. Akgol *et al.*, *Optik* **161**, 12 (2018).
31. O. Altintas *et al.*, *Mod. Phys. Lett. B* **31**, 1750274 (2017).
32. F. Dincer, M. Karaaslan and C. Sabah, *J. Electromagnetic Waves Appl.* **29**, 2492 (2015).
33. M. Bakir *et al.*, *Opt. Eng.* **54**, 097102 (2015).
34. M. Karaaslan *et al.*, *IET Microwaves, Antennas Propagation* **7**, 215 (2013).
35. F. Ding, A. Pors and S. I. Bozhevolnyi, *Rep. Progress Phys.* **81**, 026401 (2017).
36. S. Mahoney *et al.*, *PLoS One* **11**, e0152010 (2016).
37. J. Yang *et al.*, *Opt. Commun.* **402**, 518 (2017).
38. W. Ma, Y. Wen and X. Yu, *Opt. Exp.* **21**, 30724 (2013).

39. X. Ling, Z. Xiao and X. Zheng, *Plasmonic.* **13**, 175 (2018).
40. X. Ling *et al.*, *J. Mod. Opt.* **64**, 665 (2017).
41. F. Ding *et al.*, *Sci. Rep.* **6**, 39445 (2016).
42. P. Rufangura and C. Sabah, *J. Nanophoton.* **12**, 026002 (2018).
43. R. Deshmukh *et al.*, *ACS Photonic.* **5**, 2737 (2018), doi:10.1021/acsphotonics.8b00484.
44. O. A. M. Abdelraouf *et al.*, Plasmonic scattering nanostructures for efficient light trapping in flat CZTS solar cells, *Proc. SPIE 10227, Metamaterials XI*, 1022712, 16 May 2017, doi:10.1117/12.2265249, https://doi.org/10.1117/12.2265249.
45. S. Sharma *et al.*, *Renewable Sustainable Energy Rev.* **70**, 529 (2017).
46. C. Zhang *et al.*, *Opt. Mater. Exp.* **5**, 1954 (2015).
47. P. Yu *et al.*, *Adv. Opt. Mater.* 1800995 (2018), https://doi.org/10.1002/adom.201800995.
48. A. Biswas *et al.*, *Appl. Phys. Lett.* **88**, 013103 (2006).
49. M. K. Hedayati *et al.*, *Adv. Mater.* **23**, 5410 (2011).
50. M. K. Hedayati and F. E. Faupel, *M. Appl. Phys. A* **109**, 769 (2012), https://doi.org/10.1007/s00339-012-7344-1.
51. B. J. Lee *et al.*, *J. Heat Transfer* **136**, 072702 (2014).
52. S. Niu *et al.*, *Nat. Photonic.* **12**, 392 (2018).
53. T. Theivasanthi and M. Alagar, *Nano Biomed. Eng.* **5**, 11 (2014), https://doi.org/10.5101/nbe.v5i1.p11-19.
54. N. Valizade-Shahmirzadi and T. Pakizeh, *Mater. Res. Exp.* **5** (2018).

https://doi.org/10.1142/9789819818433_0021

Chapter 21

The Effect of CaO/SiO_2 and B_2O_3 on the Sintering Contraction Behaviors of CaO-B_2O_3-SiO_2 Glass-Ceramics#

Tingnan Yan, Weijun Zhang*, Haijun Mao, Xingyu Chen and Shuxin Bai

Department of Materials Science and Engineering,
College of Aerospace Science and Engineering,
National University of Defense Technology,
Changsha 410073, P. R. China
**zhwjun@nudt.edu.cn*

We investigate the effect of CaO/SiO_2 (molar ratio) and B_2O_3 content on the structure and sintering contraction behaviors of Calcium borosilicate (CaO–B_2O_3–SiO_2, CBS) glass-ceramic through the dilatometer and the Fourier transform infrared (FTIR) spectroscopy. The results show that the sintering-shrinkage of CBS glass-ceramics is promoted dramatically with the increase of CaO/SiO_2 to 1.14, and then keeps nearly constant as the CaO/SiO_2 further increases. This phenomenon is correlated with the degree of polymerization (DOP) of the Si–O network structure modulated by the CaO/SiO_2, additionally, two sintering shrinkage peaks, corresponding to two-step depolymerization, are detected in the sintering-shrinkage curves of the CBS glass-ceramic as the B_2O_3 content is elevated to 13.6 mol%. Meanwhile, the shrinking rate of CBS glass-ceramics is increased from 8×10^{-3} to 37×10^{-3} min^{-1} and the softening point is decreased from 736°C to 691°C with the increase of B_2O_3 content from 6.8 to 20.8 mol·%. This is due to the introduction of BO3 trihedral into Si–O–Si three-dimensional structure, which greatly reduces the uniformity and symmetry of the networks, inducing the decrease of the strength for the whole Si–O–Si network structures. The results obtained in this paper reveal the relationship between the glass structure and sintering behavior of the CBS glass-ceramic, which gives an avenue to improve its physical properties.

Keywords: Sintering contraction behaviors; glass structure; CaO–B_2O_3–SiO_2; LTCC.

*Corresponding author.

#To cite this article, please refer to its earlier version published in the *International Journal of Modern Physics B*, Volume 33, 1950070 (2019), DOI: 10.1142/S021797921950070X.

1. Introduction

Low-temperature co-fired ceramics (LTCC) has been widely applied in radar, wireless communication fields, antennas and IC packaging, attributed to its own modularization packaging and well co-fired compatibility with Ag or Au electrode.[1,2] The CBS glass-ceramic with outstanding mechanical and dielectric properties, such as relatively low dielectric constant ($\varepsilon_r < 6.5$), dielectric loss ($\tan\delta < 2 \times 10^{-3}$) and excellent mechanical properties, is an excellent alternative substrate material for the LTCC technology ,[3,4] and it has been intensively studied by many researchers. A great number of works have reported on the influence of Ca/Si ratio and B_2O_3 content on the dielectric properties and crystallization behavior of the CBS glass-ceramic. For instance, many researches indicate that excessive B_2O_3 will destruction the dielectric properties of CBS glass-ceramic system[5–8] and the dielectric loss ($\tan\delta$) increases and dielectric constant (ε_r) decreases gradually with the elevated Ca/Si ratio.[9] As for the crystallization behavior, reported investigations show that with the addition of B_2O_3, the major crystalline phases will transform from $CaSiO_3$ to CaB_2O_4, at the same time, the amount of α-SiO_2 crystalline phase will decrease.[10]

Sintering shrinkage is another key indicator of the glass-ceramic for its practical application as LTCC substrate materials, which has a significant effect on the mechanical strength and dielectric loss. More specifically, if the material is not dense enough, the strength of the LTCC material will decrease substantially, meanwhile, dielectric loss will also increase to a certain extent. However, there are very few works focusing on the sintering contraction behaviors, especially, modulated by the variation of Ca/Si ratio and B_2O_3 content. The experiments described here particularly investigate the sintering contraction behavior of the CBS glass-ceramic with different CaO/SiO_2 (molar ratio) and B_2O_3 content, and the sintering densification mechanism based on the relationship between the glass structure and sintering contraction behavior is revealed, aiming at giving a guidance to prepare high-quality glass-ceramic with prominent mechanical and dielectric properties.

2. Experimental

The major chemical composition of each CBS glass-ceramics used in this study is listed in Table 1. All those CBS glass-ceramics were prepared from reagent grade chemicals of CaO, SiO_2 and B_2O_3, where the CaO/SiO_2 was fixed at 0.4, 0.86, 1, 1.14, 1.2 and 1.6, respectively, and the content of

Table 1. Compositions of $CaO–B_2O_3–SiO_2$ glass-ceramics.

No.	CaO (mol·%)	B_2O_3(mol·%)	SiO_2(mol·%)	Ca/Si molar ratio	T_s(°C)
1–1	26.2	8.5	65.3	0.4	750
1–2	42.2	8.5	49.3	0.86	748
1–3	45.7	8.5	45.8	1	734
1–4	48.8	8.5	42.7	1.14	727
1–5	49.8	8.5	41.7	1.2	726
1–6	56.3	8.5	35.2	1.6	721
2–1	46.6	6.8	46.6	1	736
2–2	44.9	10.2	44.9	1	725
2–3	43.2	13.6	43.2	1	704
2–4	39.6	20.8	39.6	1	691

B_2O_3 was varied at 6.8, 10.2, 13.6 and 20.8 mol·%. In the preparation of experimental samples, the raw materials were firstly weighted according to Table 1, and then they were mechanically fully mixed. After mixed uniformly, the raw materials, compositions as shown in Table 1, were put into a platinum crucible and melted by an electric furnace at 1500°C for 2 h. The molten glass was quenched into distilled water to form cullet. Then, the cullet was dried and milled with a mixture of agate balls and ethyl alcohol to an average particle size of 2.5 μm and then oven-dried at 80°C. The dried CBS glass-ceramics powder was pelletized with 5 wt·% polyvinyl alcohol (PVA) solution, then pressed into disks (Φ 20 mm $\times$ 10 mm) under the pressure of 20 MPa for 5 min. Samples were heat treated at 450°C for 2 h to remove the PVA, then upto 850°C for 10 min, the heating rate in the whole process was 5°C/min.

Dilatometry measurements were done with prismatic samples 1 g with a cross-section of 4×5 mm^2 and 30 mm long using a dilatometer (NETZSCH DIL 402C, Germany) at the heating rate of 5 K min^{-1}. Infrared spectra were recorded with a Fourier transform infrared (FTIR) spectrometer (Thermo Fisher Scientific Nicolet 380, US) in the frequency ranges of 400–4000 cm^{-1} at room-temperature. The quenched samples were dried and milled into powders with an average particle size about 2.5 μm. Then 1.5 mg of each CBS glass-ceramic powders mixed with 150 mg KBr (reagent grade) were pressed into the thin disk for the FTIR test. The assignments of the absorption bands representing for the corresponding network structure are summarized in Table 2. The microstructure of the sintered samples was observed by SEM (scanning electron microscope, Tescan MAIA3 XMH, Czech Republic).

Table 2. Assignment of IR spectra bands of samples.

Wavenumber (cm^{-1})	Assignments	References
498–517	Bending vibration modes of Si–O–Si bonds in [SiO_4] tetrahedron	13 and 14
709–721	Bending vibration of B–O–B in [BO_3] triangles	15 and 16
798–760	Stretching vibration of O–Si-O	17
1019–1049	Asymmetric stretching vibration of Si–O–Si in [SiO_4] tetrahedron and B–O asymmetric stretching of tetrahedral [BO_4] units	16 and 18
1402–1420	Asymmetric stretching relaxation of the B–O bond of trigonal [BO_3] units	19

3. Results and Discussion

3.1. *Effect of CaO/SiO_2 on the sintering contraction behaviors of CBS glass-ceramic and its correlation to the structure*

Figure 1 shows the sintering-shrinkage curves of the CBS glass-ceramic with different CaO/SiO_2, and the content of B_2O_3 is fixed at 8.5 mol·%. It can be found that the maximum line shrinkage of CBS glass-ceramic is strongly associated with its CaO/SiO_2. As described in Fig. 1(a), when the CaO/SiO_2 is below 0.86, the maximum line shrinkage of the material is less than 5%, and the densification of the CBS glass can be hardly achieved at 800°C. However, as the CaO/SiO_2 increased to 1, a sharp increase of the shrinkage takes place, reaching 19.9%. In addition, the variation of the line shrinkage of the CBS becomes unremarkable with the further increase of CaO/SiO_2 from 1.14 to 1.6, corresponding rates are 21.8% for 1.14, 22.5% for 1.2, and 25.3% for 1.6, respectively. Further analysis of the data reveals that the softening points of the CBS glass-ceramics show a decreasing trend with the increased CaO/SiO_2, but the decrease is not obvious. And the softening points are about 750°C, 748°C, 737°C, 722°C and 721°C for the samples with the CaO/SiO_2 of 0.4, 0.86, 1, 1.14, 1.2 and 1.6, respectively.

The curves shown in Fig. 1(b) are obtained by the derivative of time t in Fig. 1(a), we can see that the glass shrinking speed is very slow before the CaO/SiO_2 increases to 1, and no obvious shrinking peak is observed. Subsequently, the shrinking speed increases significantly as the CaO/SiO_2 reaches 1 and the maximum shrinking speed is around 770°C. When the CaO/SiO_2 rises to 1.14, the glass shrinking peak moves to about 750°C, and the shrinking speed has a certain improvement compared with sample 1–3.

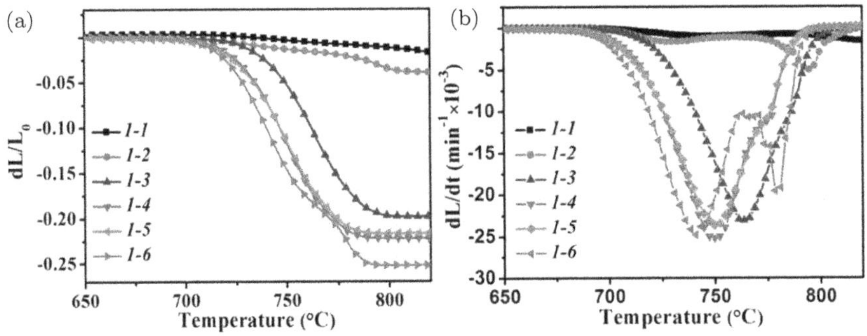

Fig. 1. (Color online) Dilatometric analyses results for CBS glass-ceramics at different CaO/SiO_2: (a) dilatometric curves (dL/L_0) and (b) the derivatives of Dilatometric curves (dL/dt).

However, further increase of the CaO/SiO_2 leads to less variation of the shrinking speed of CBS glass-ceramic.

Generally, it can be considered that the sintering mechanism of the LTCC systems is involved with the particle rearrangement under capillary force associated with viscous flow of melt. The kinetics for the initial phase of sintering is described as follows[11]:

$$\Delta L/L = (3\gamma/4r\eta) \times t, \tag{1}$$

where η is the viscosity of the glass, γ is the surface tension, r is radius of particle and $\Delta L/L$ is the shrinking percentage. Given that the surface tension of silicate glass is a fixed value of about 0.3 to 0.35,[12] the sintering-shrinkage rate and the viscosity of the glass are in inverse proportion for the CBS glass-ceramics.

It is thought that the depolymerization of the network structure will lead to a lower viscosity of the glass. In our work, as the CaO/SiO_2 increases, the amount of the free oxygen ions O^{2-} increases owing to the intensified dissociation of CaO. As a result, the free oxygen ions O^{2-} will react with the bridged oxygen O^0 in the silicates, and then decrease the agglomerate Si–O structure within the CBS glass-ceramics.[9] Thus, the viscosity of CBS glass-ceramics should be decreased with the increase in CaO/SiO_2, then inducing a high sintering-shrinkage rate. However, when the CaO/SiO_2 is elevated to 1.14, the depolymerization effect is not obvious even though the number of free oxygen ions O^{2-} is continually expanded, as many existing agglomerate Si–O structures have already been destructed into smaller and simpler network units.

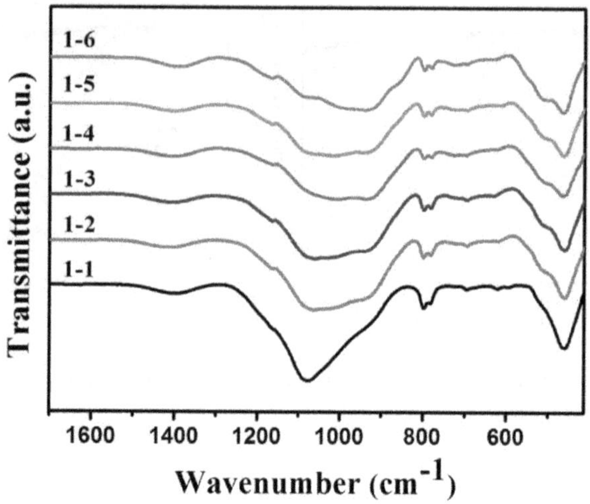

Fig. 2. (Color online) Effect of CaO/SiO_2 on the results of FTIR spectra of the quenched samples.

The FTIR spectra results of the quenched samples with varied CaO/SiO_2 are shown in Fig. 2, where the assignments of each band are listed in Table 2. When the CaO/SiO_2 is below 1.2, it could be listed that the absorbance peaks of the O–Si–O bond Stretching vibration are in the region of 760–798 cm^{-1}, the asymmetric stretching vibration of Si–O–Si in SiO_4 tetrahedron of 1019–1049 $cm^{-1}$and B–O asymmetric stretching in tetrahedral BO_4 units of 1019–1049 cm^{-1} become less pronounced with the augment of CaO/SiO_2, this is because of more free oxygen ions O^{2-} reacting with the bridged oxygen O^0 of Si–O–Si within the CBS glass-ceramics structure, which results in the depolymerization of the Si–O–Si bond. At the same time, further increase of the CaO/SiO_2 from 1.14 to 1.6 leads to little variation of the FTIR spectra, owing to the fact that many existing agglomerate Si–O structures have already been destructed into smaller and simpler network units. The schematic illustration for the above process is shown in Fig. 3.

The microstructures of the samples with different CaO/SiO_2 detected by SEM are presented in Fig. 4. All the samples are sintered at 850°C for 10 min at a heating rate of 5 K min^{-1}. The SEM images show that samples 1–1 and 1–2 contain many holes, and most holes are connected. The reason is that when the CaO/SiO_2 is less than 0.86, the softening point of the CBS glass-ceramics is as high as 748°C, and the viscosity is

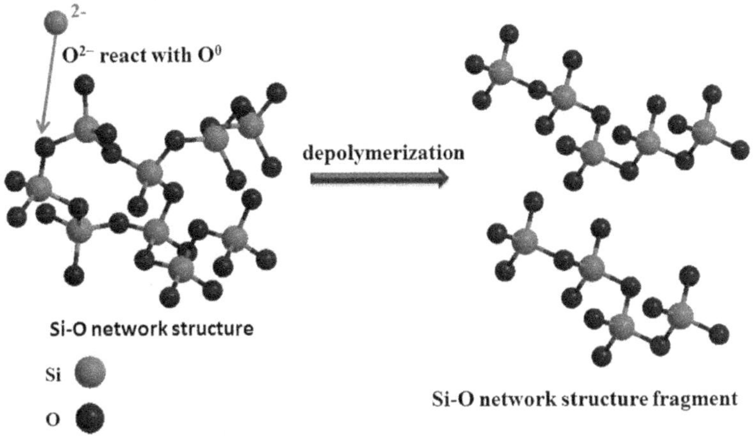

Fig. 3. (Color online) The schematic illustration for increase CaO/SiO_2 in CBS glass-ceramics.

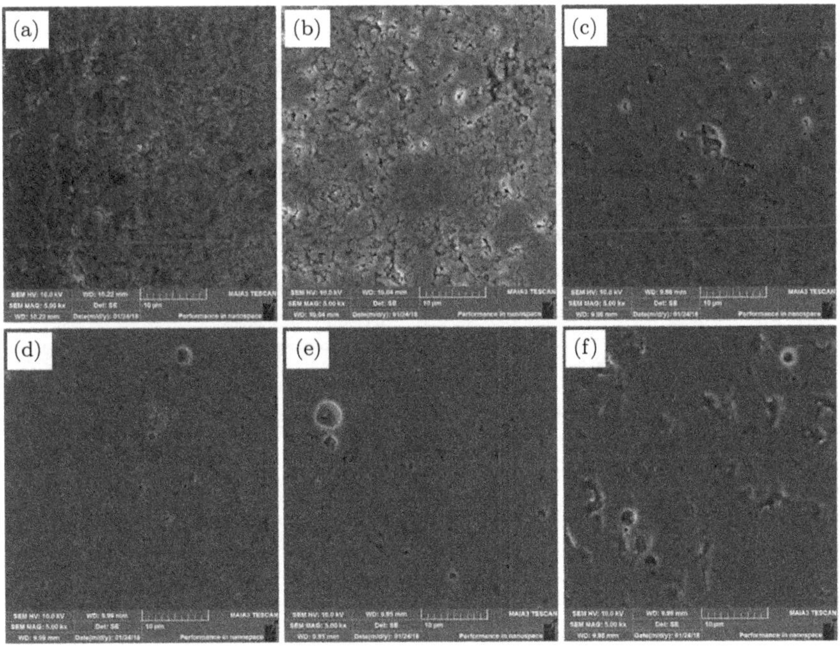

Fig. 4. SEM images of polished CBS glass-ceramics (a) 1–1; (b) 1–2; (c) 1–3; (d) 1–4; (e) 1–5; (f) 1–6, sintering in 850°C.

still very large at the final sintering temperature of 850°C, resulting in a small sintering-shrinkage rate and porous microstructure of samples 1–1 and 1–2. Comparing samples 1–3 and 1–4, we can find that the number of holes in 1–4 is greatly reduced. When the CaO/SiO_2 is greater than 1 and less than 1.14, the softening point of CBS glass-ceramics and sintering viscosity decrease with the increase of CaO/SiO_2, and then the density of CBS glass-ceramics is greatly increased, which result in more densification of samples 1–4 than 1–3. As the CaO/SiO_2 continues to increase, the size of holes in 1–4, 1–5 and 1–6 successively increases. This is because when the CaO/SiO_2 is greater than 1.14, the softening point of CBS glass-ceramics and the sintering-shrinkage rate are no longer changed significantly. However, with the increase of the CaO/SiO_2, a large number of Ca^{2+} ions will lower the glass crystallization temperature, which is detrimental to the densification of 1–5 and 1–6.

3.2. *Effect of B_2O_3 on the sintering contraction behaviors CBS glass-ceramic and its correlation to the structure*

Figure 5 shows the sintering-shrinkage curves of the CBS glass-ceramic with different B_2O_3 content and CaO/SiO_2 fixed at 1. As described in Fig. 5(a), the maximum line shrinkages are 17.4%, 23.7%, 22.8% and 23.6% for the samples with the B_2O_3 content of 6.8, 10.2, 13.6 and 20.8 mol·%, respectively. Further analysis of the data reveals that the softening points of the CBS glass-ceramics show a decreasing trend with the increased B_2O_3 content, and the softening points are about 736°C, 725°C, 704°C and 691°C

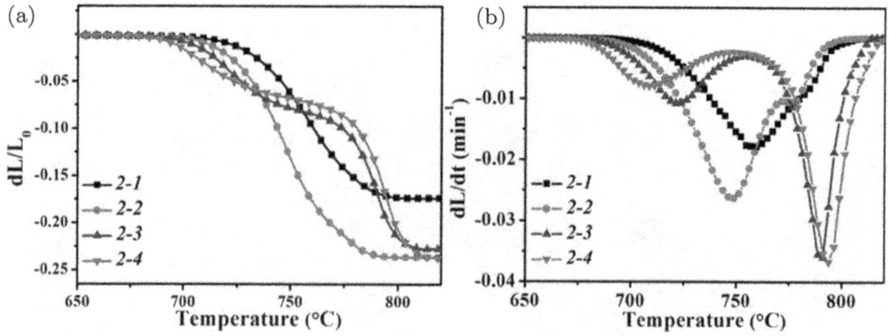

Fig. 5. (Color online) Dilatometric analyses results for CBS glass-ceramics at different B_2O_3 content: (a) dilatometric curves (dL/L_0) and (b) the derivatives of dilatometric curves (dL/dt).

for the samples with the B_2O_3 content of 6.8, 10.2, 13.6 and 20.8 mol·%, respectively. The fusing point of B_2O_3 is 450°C, and the addition of B_2O_3 will lead to the combination of B_2O_3 with other oxides to form eutectics with low fusing points, which substantially lower the softening point of glass.[20]

Figure 5(b) shows that when the content of B_2O_3 is below 10.2 mol·%, the sintering shrinking speed of glass increases with the increase of B_2O_3 content, and the maximum sintering shrinking speed is around 760°C. When the content of B_2O_3 rises to 13.6 mol·%, two different sintering shrinking peaks appear in the sintering-shrinkage curve. The sintering speed corresponding to the first sintering shrinking peak is much smaller than that of the second one. Compared with the sintering-shrinking peak of sample 2–2, the first sintering peaks of 2–3 and 2–4 appear earlier, but the intensity of the sintering shrinking peak is lower. The second sintering shrinking peak appears at a higher temperature, and the intensity of the peak is much larger than the peak corresponding to sample 2–2.

It is reported that in the vitreous B_2O_3, about 80% of the boron atoms are presented in the B_3O_6 boroxol rings, which are interconnected by independent BO_3 groups.[20] In CBS glass-ceramics, the existing free oxygen ions O^{2-} will interact with the bridged oxygen O^0 between the BO_4 tetrahedral structural unit and the BO_3 trihedral structural unit to depolymerize the diborate structural units.[20]

The FTIR spectra data of aforementioned samples are shown in Fig. 6. It can be listed that the asymmetric stretching relaxation of the B–O bond of trigonal BO_3 units in the region of 1402–1420 cm^{-1}, and bending vibration of B–O–B of BO_3 triangles in the region of 709–721 cm^{-1} become more pronounced with the increase of B_2O_3 for the reason that free oxygen ions O^{2-} is added to the diborate structural unit, depolymerizing it into a BO_4 tetrahedral structural unit and BO_3 trihedral structural unit.

Furthermore, the absorbance peaks of the O–Si–O bond in the region of 498–517 cm^{-1} (bending vibration modes) and 760–798 cm^{-1} (stretching vibration modes), become less pronounced with the augment of B_2O_3. As mentioned above, free oxygen ions O^{2-} can interact with the bridged oxygen in diborate structural units with the increase of B_2O_3. It can be concluded that the BO_3 trihedral is introduced into Si–O–Si three-dimensional structure, which will greatly reduce the symmetry of the current structure. Additionally, the borate structure preferentially connects to the near Si atoms and is introduced into the silicate networks, which can lower the uniformity of the networks. Thus, the whole Si–O–Si structure

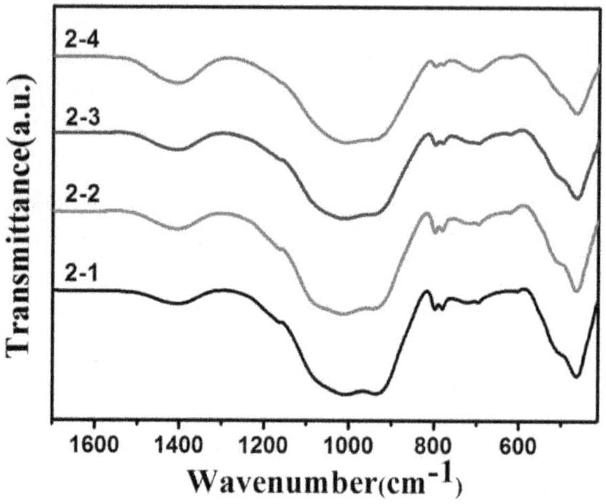

Fig. 6. (Color online) Effect of B_2O_3 on the results of FTIR spectra of the quenched samples.

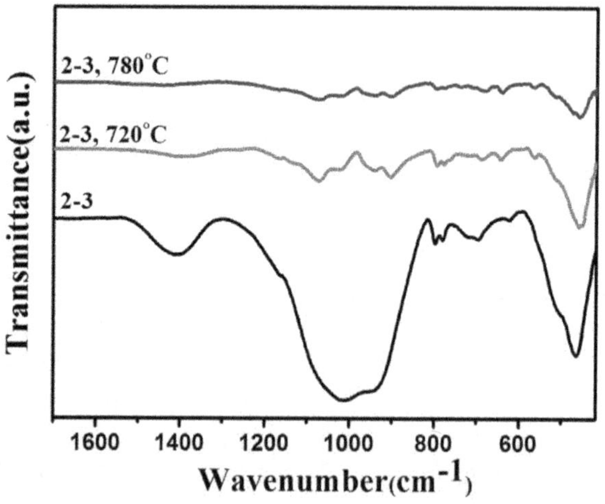

Fig. 7. (Color online) Effect of sintering temperature on the results of FTIR spectra of the 2–3 CBS glass-ceramics.

is destroyed and the viscosity of CBS glass-ceramics decreases with the addition of B_2O_3.

Figure 7 shows the FTIR spectrum of sample 2–3 fired at 720°C and 780°C, it can be found that the asymmetric stretching relaxation of the

B–O bond of trigonal BO$_3$ units in the region of 1402–1420 cm^{-1} and the Bending vibration of B–O–B in BO$_3$ triangles in the region of 709–721 cm^{-1} almost disappear with the temperature goes upto 720°C, due to the utterly destroyed BO$_3$ units in the network. In the meantime, it can be noted that the absorbance peaks in the region of 498–517 cm^{-1} and 760–798 cm^{-1} become indistinctive with the temperature goes upto 780°C, which indicates the fracture of Si–O network. Excessive B$_2$O$_3$ content leads to gradual breakdown of the network structure, which explains the two different sintering peaks in Fig. 5 (b) for samples 2–3 and 2–4. The schematic illustration for the above process is shown in Fig. 8.

The microstructures of the samples with different B$_2$O$_3$ content observed by SEM are presented in Fig. 9, all the samples are sintered at 850°C for 10 min. Comparing samples 2–1 and 2–2, we can find that the number of holes in 2–2 is greatly reduced as shown in the SEM images. When the B$_2$O$_3$ content is less than 10.2 mol·%, the softening point of CBS glass-ceramics and sintering viscosity decreases with the increase of B$_2$O$_3$ content, and then the density of CBS glass-ceramics is greatly increased. As B$_2$O$_3$ content continues to increase, the number of holes in 2–2, 2–3 and 2–4 increases successively. This is because when the content of B$_2$O$_3$ is greater than 13.6 mol·%, the sintering of CBS glass-ceramics is divided

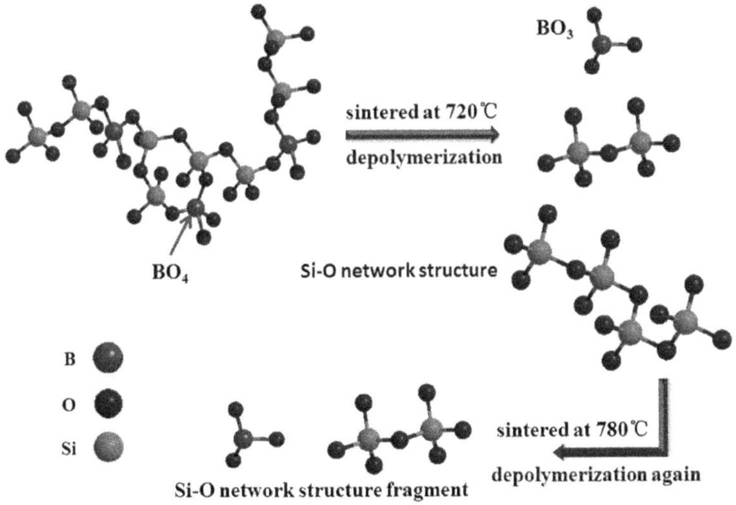

Fig. 8. (Color online) The schematic illustration for increase CaO/SiO$_2$ in CBS glass-ceramics.

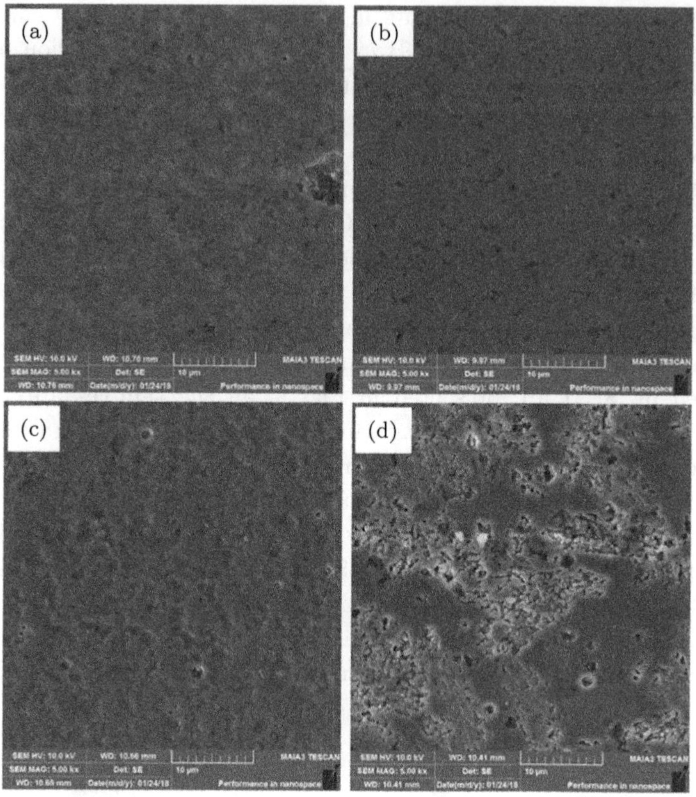

Fig. 9. SEM images of polished CBS glass-ceramics (a) 2–1; (b) 2–2; (c) 2–3; (d) 2–4, sintering in 850°C.

into two steps. The first step is related to the sintering peak at a low sintering temperature about 710°C, but the sintering-shrinkage rate is very slow. Although the second sintering peak is associated with a very fast sintering shrinking speed, the sintering shrinking peak is extremely close to the crystallization temperature, leading to an extremely short sintering time, which is detrimental to the densification of the glass-ceramic. In conclusion, with the increase of B_2O_3 content, the softening point of CBS glass-ceramics may be reduced to a certain extent, but when the content of B_2O_3 is greater than 13.6 mol·%, the fast sintering process is too short, and the glass is difficult to be dense, which results in more porous structures in samples 2–2, 2–3 and 2–4.

4. Conclusions

The effect of CaO/SiO_2 and B_2O_3 content sintering contraction behaviors of CBS glass-ceramics related to the glass structure is researched in this paper, and the conclusions are summarized as follows:

(1) With the increase of CaO/SiO_2, the sintering shrinking speed of CBS glass-ceramics gradually increases, and then keeps nearly constant with the elevated CaO/SiO_2. This can be explained by the depolymerization of the Si–O network structure with the addition of CaO/SiO_2, however, the further depolymerization is less pronounced with the CaO/SiO_2 being greater than 1.14.

(2) The sintering shrinking rate increases obviously with the addition of B_2O_3 content because the introduction of BO_3 trihedral into Si–O–Si three-dimensional structures greatly reduces the uniformity and symmetry of the networks, which sharply decreases the strength of the whole Si–O network structures. The result of sintering-shrinkage curves indicates that further addition of B_2O_3 (>13.6 mol·%) leads to two sintering peaks for the CBS glass-ceramics, corresponding to two-step depolymerization of the CBS glass-ceramic.

References

1. J. Zhou, *J. Adv. Ceram.* **1**, 89 (2012).
2. L. J. Golonka, *Bull. Pol. Acad. Sci. Tech. Sci.* **54**, 221 (2006).
3. C. Chiang and S. F. Wang, *J. Alloys Compd.* **461**, 612 (2008).
4. H. K. Zhu *et al.*, *J. Mater. Sci., Mater. Electron.* **20**, 1135 (2009).
5. C. C. Chiang *et al.*, *Ceram. Int.* **34**, 599 (2008).
6. S. F. Wang *et al.*, *J. Alloys Compd.* **498**, 211 (2010).
7. S. H. Wang and H. P. Zhou, *Mater. Sci. Eng. B* **99**, 597 (2003).
8. H. B. Shao, T. W. Wang and Q. T. Zhang, *J. Alloys Compd.* **484**, 2 (2009).
9. X. H. Zhou *et al.*, *J. Mater. Sci. Mater. Electron.* **20**, 262 (2009).
10. A. Q. Jia *et al.*, *Key Eng. Mater.* **680**, 301 (2015).
11. W. D. Kingery, H. K. Bowen and D. R. Uhlmann, *Introduction to Ceramics* (Wiley & Sons, New York, 1976), p. 492.
12. C. F. Yang and C. N. Cheng, *Ceram. Int.* **25**, 383 (1999).
13. S. V. Stefanovsky, K. M. Fox and J. C. Marra, *Mater. Res. Soc.* **1518**, 53 (2013).
14. K. H. S. Shaaban *et al.*, *Silicon* **10**, 1 (2018).
15. H. Doweidar and Y. B. Saddeek, *J. Noncryst. Solids* **356**, 1452 (2010).
16. K. H. S. Shaaban *et al.*, *Appl. Phys. A* **123**, 457 (2017).

17. P. F. Wei *et al.*, *J. Central South Univ. Technol.* **18**, 1359 (2011).
18. J. P. Wang, J. H. Cheng and P. Lu, *J. Wuhan Univ. Technol.-Mater.* **23**, 419 (2008), doi:10.1007/S11595-007-3419-9.
19. E. I. Kamitsos *et al.*, *J. Noncryst. Solids* **126**, 52 (1990).
20. I. Kashif *et al.*, *Physica B* **403**, 3903 (2008).

https://doi.org/10.1142/9789819818433_0022

Chapter 22

Polymorphic Transformations and Thermal Expansion of Some Modifications in $Ag_{1.5}Cu_{0.5}Se$ and $Ag_{0.4}Cu_{1.6}Se$[#]

Y. I. Aliyev[*,†], Y. G. Asadov[*], A. O. Dashdemirov[†], R. D. Aliyeva[*], T. G. Naghiyev[*] and S. H. Jabarov[*,†,‡]

[*] *Institute of Physics, Azerbaijan National Academy of Sciences, Baku AZ-1143, Azerbaijan*
[†] *Department of Physics and Technology, Azerbaijan State Pedagogical University, Baku AZ-1000, Azerbaijan*
[‡] *sakin@jinr.ru*

The $Ag_{1.5}Cu_{0.5}Se$ and $Ag_{0.4}Cu_{1.6}Se$ compounds have been synthesized and grown as single crystals. High-temperature X-ray diffraction method was used to study polymorphic transformations. It is shown that the $Ag_{1.5}Cu_{0.5}Se$ crystals of high-temperature FCC modification are decomposed into Ag_2Se and AgCuSe when the temperature decreases below $T = 488$ K and $Ag_{0.4}Cu_{1.6}Se$ is decomposed into Cu_2Se and AgCuSe when the temperature decreases below $T = 540$ K. Transformations in both compounds are reversible. Crystalline parameters are obtained and the temperature dependence of the lattice parameters for each phase is built.

Keywords: Crystal structure; XRD; phase transition; semiconductors.

1. Introduction

The structural properties of copper containing chalkogenides which have a semiconductor properties are widely investigated.[1–5] The main reason for great interest in these materials in crystallography is due to different phase transitions that occur in high-temperature range.[6–10] Therefore, it is very important to study the structure phases of these compounds in high-temperature range.

[‡]Corresponding author.

[#]To cite this article, please refer to its earlier version published in the *International Journal of Modern Physics B*, Volume 33, 1950271 (2019), DOI: 10.1142/S0217979219502710.

AgCuSe compound is formed at Cu_2Se:Ag_2Se = 1:1 ratio on peritectic reaction at temperature $T = 1033$ K.[11,12] In Ref. 13, AgCuSe crystal lattice is determined as tetragonal with lattice parameters $a = 4.075$ Å, $b = 4.07$ Å, $c = 6.29$ Å, $Z = 2$, sp.gr. D^7_{4h} – Pnmm, $\rho = 7.91$ g/sm^3. Later, in Ref. 13, AgCuSe crystal lattice is determined as orthorhombic with parameters $a = 4.105$ Å, $b = 4.07 \times 5 = 20.35$ Å, which is the superstructure of tetragonal lattice. Parameter b of orthorhombic lattice equals five parameters of b of tetragonal lattice. According to data of Ref. 14, Ag atoms in AgCuSe orthorhombic structure are in planes perpendicular to c-axis. Around each of them, there are four Ag atoms at a distance of $d = 2.96$ Å and six Se atoms at distances of $d = 2.67$ Å (4Se), 3.59 Å (1Se) and 3.64 Å (1Se). Se atoms form stretched tetrahedrons with Cu atoms in the center. Distance $d_{\mathrm{Se-Se}} = 3.03$ Å, $d_{\mathrm{Cu-Se}} = 2.06$–2.50 Å and minimum spacing $d_{\mathrm{Cu-Ag}} = 2.98$ Å.

In Refs. 15–20, it is shown that orthorhombic modification at $T =$ 504 K transforms into high-temperature FCC modification with unit cell parameter of $a = 6.082$ Å. Transformation is reversible and occurs on single crystal–single crystal type.

In this work, structural variations in $Ag_{1.5}Cu_{0.5}Se$ and $Ag_{0.4}Cu_{1.6}Se$ compounds have been studied by using high-temperature roentgenography method, where numbers of Ag and Cu cations in AgCuSe are changed.

2. Experimental

The $Ag_{1.5}Cu_{0.5}Se$ and $Ag_{0.4}Cu_{1.6}Se$ compounds have been synthesized and grown as single crystals. The following primary components Ag and Cu 99.999%, Se– of "B5" were used for $Ag_{1.5}Cu_{0.5}Se$ and $Ag_{0.4}Cu_{1.6}Se$ synthesis.

The synthesis conditions did not differ from AgCuSe synthesis conditions.[15–17] X-ray phase analysis showed that both the compounds at room-temperature are two-phased and $Ag_{1.5}Cu_{0.5}Se$ consists of AgCuSe and Ag_2Se, but $Ag_{0.4}Cu_{1.6}Se$ consists of AgCuSe and Cu_2Se, i.e., in both the compounds, the orthorhombic phase consists of two structural phases and the high-temperature cubic phase consists of one structural phase. Therefore, it was impossible to grow single crystals from low-temperature phases.

The obtained ingot in crushed form was placed in the pointed ampoule, which provided the selection of high-temperature germ modification.

The ampoules were pumped out until pressure of 10^{-3} Pa was placed in the upper area of furnace and heated upto $T = 1400$ K, then released through the temperature gradient zone with the speed of 2 mm/h. It is important to note that high-temperature modification crystals are formed and grown when ampoule starts passing below the melting point. The low-temperature modification crystals are formed and grown from high-temperature modification crystals with decreasing temperature. During the growth of new modification single crystals within single crystal matrix, the density difference is one of the main factors affecting to single crystallization of new modification. According to density difference of modifications, polymorphic substances can be divided into the following three groups: (a) density of growing modification ρ_p is less than density of matrix crystal ρ_m, i.e., $\rho_p < \rho_m$; (b) density of growing modification is more than density of matrix crystal, i.e., $\rho_p > \rho_m$; (c) densities of growing and matrix crystals are either equal or vary insignificantly, i.e., $\rho_p = \rho_m$, $\rho_p \geq \rho_m$ or $\rho_p \leq \rho_m$.

During the transformation process at $\rho_p < \rho_m$, the matrix crystal is deformed and very often decrepitated and each induced break becomes source of formation of a number of crystallization centers. As a result, high-temperature modification crystal turns into low-temperature phase polycrystal. In case of $\rho_p > \rho_m$, the contact between growing crystal and crystal-medium weakens and the distance of the surface phase separation becomes more than the interatomic distance. The growth of new phase crystal is stopped under these conditions. The transformation process continues due to the formation of new germs. In this case, the single crystal turns into polycrystal. The single crystal cannot be grown from such polycrystal. Two types of transformation are usually observed in case of $\rho_p \geq \rho_m$, $\rho_p \leq \rho_m$ and $\rho_p = \rho_m$: single crystal–single crystal and single crystal–polycrystal. Single crystal–polycrystal transformation occurs due to the defectiveness of the matrix crystal and accompanied by the formation of multiple centers of crystallization.

The single crystal of required modification can be obtained from such a polycrystal at certain number of transformations.

High-temperature studies were carried out on X-ray diffractometer DRON-3M (CuK_α radiation, $\lambda_\alpha = 1.5418$ Å and Ni-filter) with add-on device URVT-2000, ensuring vacuum of 10^{-2} Pa. Angle-resolved record constituted ~0.1°. Reflection angle determination error did not exceed $\Delta\theta = \pm 0.02°$.

3. Results and Discussion

3.1. $Ag_{1.5}Cu_{0.5}Se$

Figure 1 depicts the X-ray diffraction spectra of $Ag_{1.5}Cu_{0.5}Se$ measured at different temperatures. There were 27 diffraction reflections recorded within 1 Å $\leq d_{hkl} \leq$ 3.5 Å angular spacing at room-temperature from $Ag_{1.5}Cu_{0.5}Se$ sample in the form of plate with $4\times5\times1$ mm dimensions. For accurate indication of $Ag_{1.5}Cu_{0.5}Se$ diffraction data, experimental values of planar spacing d_i were compared with planar spacing designed on the basis of AgCuSe, Ag_2Se and Cu_2Se lattice parameters of low-temperature modifications. The experimental values of the interplanar distances d_{hkl} in the $Ag_{1.5}Cu_{0.5}Se$ crystal, with the exception of $d = 2.4812$ Å and $d = 1.8625$ Å, satisfactorily coincide with the calculated values of interplanar distances for Ag_2Se crystals with orthorhombic lattice parameters $a = 4.333$ Å, $b = 7.062$ Å, $c = 7.764$ Å. The two above-indicated interplanar distances as well as number of others are indicated also on the basis of AgCuSe lattice parameters of orthorhombic modification, where $a = 4.105$ Å, $b = 20.350$ Å, $c = 6.310$ Å.

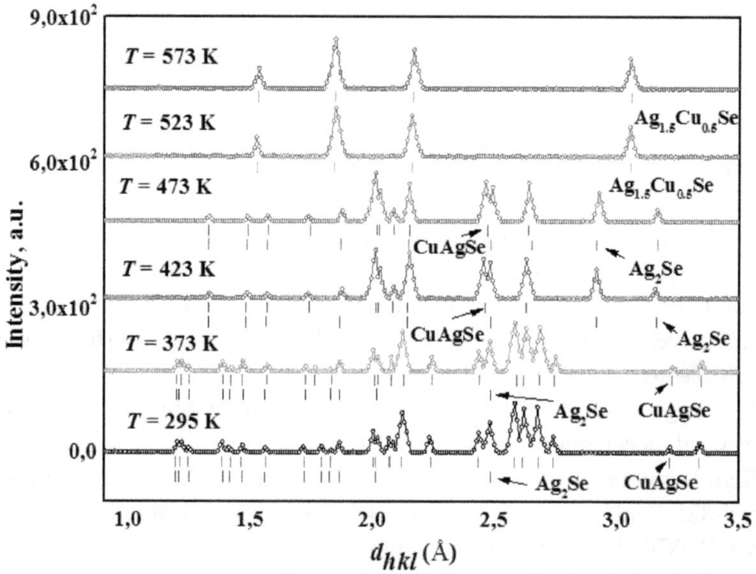

Fig. 1. (Color online) X-ray diffraction spectra of $Ag_{1.5}Cu_{0.5}Se$ measured at different temperature.

No change occurs in diffraction patterns recorded at $T = 293$–423 K temperature range. There were 14 reflections recorded, which also remained to be two phases at 423 K in previous angular range. At temperature $T = 488$ K, all reflections disappeared and four new reflections with indices (200), (220), (311) and (400) are recorded in the previous angular range, which correspond to high-temperature FCC modification with lattice parameter of $a = 6.107$ Å.

Temperature diffraction records were used to calculate lattice parameters of existing modifications, which are given in Table 1 and graphically in Fig. 2. As can be seen from Fig. 2, the values of parameters a and c of Ag_2Se crystal and parameter a of AgCuSe crystal increase linearly depending on temperature, but parameter b of Ag_2Se phase and parameters b and c of AgCuSe phase are deviated from linearity.

Temperature dependence of lattice parameters was used to calculate thermal expansion factors of both orthorhombic and FCC phases of $Ag_{1.5}Cu_{0.5}Se$, which are given in Table 2. If assuming that α' is the coefficient of thermal expansion of Ag_2Se and α'' is for AgCuSe, then it is evident from the table that $\alpha'_{[100]} \approx \alpha''_{[100]}$, $\alpha'_{[010]} < \alpha''_{[010]}$ and $\alpha'_{[001]} > \alpha''_{[001]}$. With such thermal expansion, at 488 K, both compounds simultaneously transform into single cubic phase. Transformation is reversible.

3.2. $Ag_{0.4}Cu_{1.6}Se$

The sample size was $2 \times 4 \times 6$ mm^3 which is cut off from $Ag_{0.4}Cu_{1.6}Se$ compound at room-temperature. Eighteen clear diffraction reflections were recorded from such sample in $10° \leq \Delta\theta \leq 100°$ angular range (Fig. 3). Out of 18 reflections, 15 are indicated on the basis of parameters of

Table 1. Temperature dependence of $Ag_{1.5}Cu_{0.5}Se$ lattice parameters.

T_{exp}, K	Modification	Lattice parameters (Å)			Z	sp.gr.	V, Å^3	ρ, g/sm^3
		a	b	c				
293	Ag_2Se-orthorhombic	4.333	7.062	7.764	4	$P2_12_12_1$	237.576	8.237
	AgCuSe-orthorhombic	4.105	20.350	6.310	10	P4/nmm	527.117	7.885
373	Ag_2Se-orthorhombic	4.340	7.084	7.787	4	$P2_12_12_1$	239.408	8.175
	AgCuSe-orthorhombic	4.110	20.481	6.307	10	P4/nmm	530.904	7.829
423	Ag_2Se-orthorhombic	4.354	7.068	7.869	4	$P2_12_12_1$	242.161	8.119
	AgCuSe-orthorhombic	4.085	20.797	6.317	10	P4/nmm	536.665	7.744
473	Ag_2Se-orthorhombic	4.364	7.078	7.900	4	$P2_12_12_1$	244.013	8.019
	AgCuSe-orthorhombic	4.090	20.793	6.330	10	P4/nmm	538.325	7.721
523	$Ag_{1.5}Cu_{0.5}Se$-FCC	6.107	6.107	6.107	4	Fm3m	227.753	7.950
573	$Ag_{1.5}Cu_{0.5}Se$-FCC	6.118	6.118	6.118	4	Fm3m	228.936	7.900

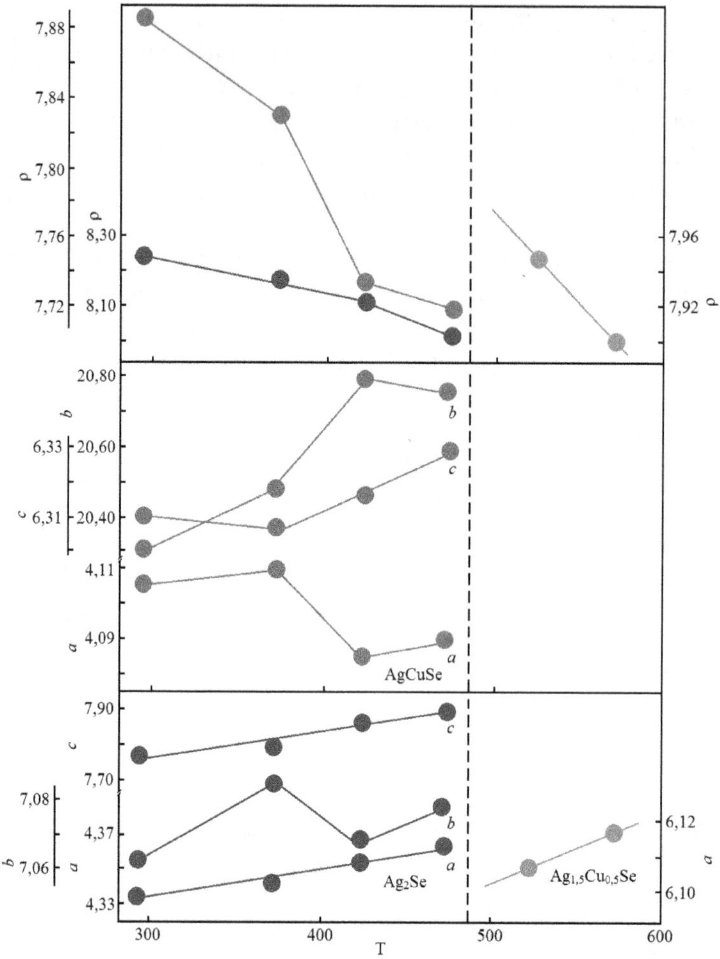

Fig. 2. (Color online) Temperature (K) dependences of lattice parameters and density of modifications $Ag_{1.5}Cu_{0.5}Se$: red line — parameters a, b, c (Å) of crystal lattice and density (g/cm) AgCuSe; blue line — parameters a, b, c (Å) of crystal lattice and density (g/cm) Ag_2Se; green line — parameter a (Å) of lattice and density (g/cm) of FCC modification $Ag_{1.5}Cu_{0.5}Se$.

orthorhombic modification AgCuSe, (need for modification here not clear) and remaining three and eleven out of fifteen are also indicated on the basis of parameters of orthorhombic modification Cu_2Se with lattice parameters $a = 4.137$ Å, $b = 7.022$ Å, $c = 20.390$ Å, $Z = 12$, sp.gr. $P222_1$. After diffraction reflections at room-temperature are recorded with no changes

Table 2. Thermal expansion factors of $Ag_{1.5}Cu_{0.5}Se$ crystal ($\times 10^6$ K^{-1}).

Modification	T, K	$\alpha_{[100]}$	$\alpha_{[010]}$	$\alpha_{[001]}$	$\bar{\alpha} = \frac{\alpha_{[100]}+\alpha_{[010]}+\alpha_{[001]}}{3}$	$\beta = \alpha_{[100]} + \alpha_{[010]} + \alpha_{[001]}$
Ag_2Se-orthorhombic	295–373	19.04	39.65	36.71	31.80	95.40
	295–423	37.28	5.99	103.83	49.03	147.10
	295–473	40.13	12.67	96.96	49.92	149.76
AgCuSe-orthorhombic	295–373	15.23	80.34	−6.14	29.81	89.43
	295–423	37.48	169.08	8.53	71.70	215.09
	295–473	33.56	120.88	17.26	57.23	171.70
$Ag_{1.5}Cu_{0.5}$Se-FCC	523–573	37.99	37.99	37.99	37.99	113.97

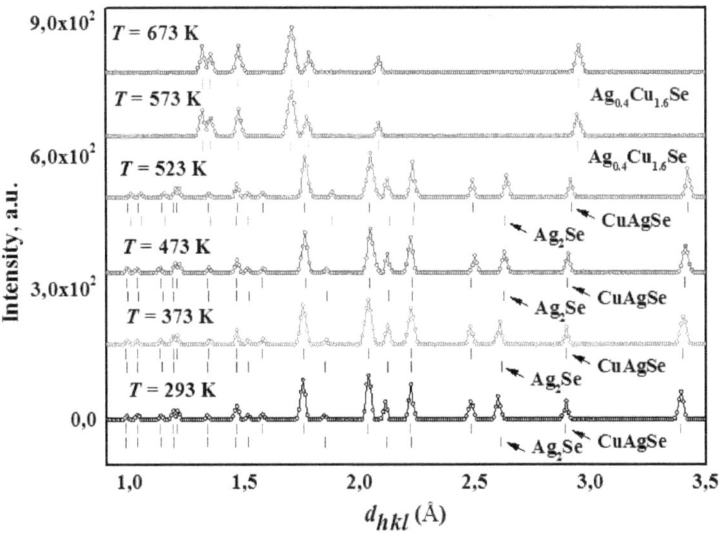

Fig. 3. (Color online) X-ray diffraction spectra of $Ag_{0.4}Cu_{1.6}Se$ measured at different temperature.

in sample orientation, furnace was turned on and diffraction records were made after every T = 50 K. Temperature of sample prior to start of each record was held constant for 50 min. No changes occurred within $T = 293$–523 K temperature range in numbers and intensity of diffraction reflections. At T = 573 K, all diffraction reflections disappear and seven new reflections are recorded in the previous angular range, which belong to high-temperature FCC modification with unit cell parameters $a = 5.826$ Å, $Z = 4$, sp.gr. Fm3m, $\rho = 7.258$ g/sm^3. Each diffraction record was used

Table 3. Temperature dependence of $Ag_{0.4}Cu_{1.6}Se$ lattice parameters.

T_{exp}, K	Modification	Lattice parameters (Å)			Z	sp.gr.	V, Å^3	ρ, g/sm^3
		a	b	c				
293	Cu_2Se-orthorhombic	4.137	7.022	20.390	12	$P222_1$	592.330	6.930
	AgCuSe-orthorhombic	4.102	20.362	6.292	10	P4/nmm	525.539	7.909
373	Cu_2Se-orthorhombic	4.139	7.029	20.434	12	$P222_1$	594.487	6.904
	AgCuSe-orthorhombic	4.106	20.389	6.295	10	P4/nmm	527.000	7.886
473	Cu_2Se-orthorhombic	4.147	7.042	20.572	12	$P222_1$	600.768	6.832
	AgCuSe-orthorhombic	4.114	20.421	6.300	10	P4/nmm	529.276	7.851
523	Cu_2Se-orthorhombic	4.154	7.052	20.666	12	$P222_1$	605.390	6.780
	AgCuSe-orthorhombic	4.121	20.432	6.309	10	P4/nmm	531.220	7.822
573	$Ag_{0.4}Cu_{1.4}Se$-FCC	5.826	5.826	5.826	4	Fm3m	197.645	7.258
673	$Ag_{0.4}Cu_{1.4}Se$-FCC	5.906	5.906	5.906	4	Fm3m	206.006	7.211

to estimate/calculate lattice parameters of existing phases, which are given in Table 3.

Equilibrium temperature between modifications equals $T = 540$ K. Transformations are reversible and high-temperature cubic modification below equilibrium temperature is decomposed into AgCuSe and Cu_2Se with both having orthorhombic structure. Temperature dependences of lattice parameters of both orthorhombic and FCC modifications of $Ag_{0.4}Cu_{1.6}Se$ are given in Fig. 4. As can be seen from Fig. 4, parameters a and b AgCuSe at $T = 473$ K deviate from linearity.

Dependence of the parameter a of the Cu_2Se phase is observed upto 373 K temperature, and this parameter increases linearly with further increase in temperature to the phase transition point. Deviation of lattice parameters from straightforwardness of both orthorhombic phases is mainly associated with redistribution of two kinds of cations. Temperature dependence of density of existing modifications is also given in Fig. 2. Existence of difference between phase densities is one of the reasons of instability of compounds considered.

Thermal expansion factors estimated/calculated from temperature dependence of lattice parameters of existing phases in $Ag_{0.4}Cu_{1.6}Se$ composition are given in Table 4. As is evident from Table 4, there is a strong anisotropy in thermal expansion factors in both orthorhombic compounds on main crystallographic directions. It is also one of the main reasons for temperature instability of crystal structure.

In conclusion, we note that the Ag cations are larger than Cu in the $Ag_{1.5}Cu_{0.5}Se$ crystal, therefore, the single high-temperature FCC modification below the equilibrium temperature ($T_0 = 488$ K) decomposes

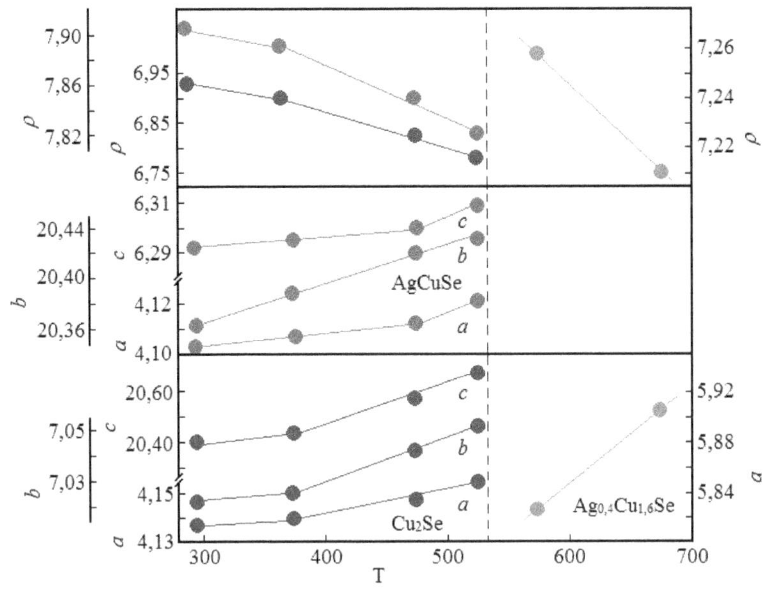

Fig. 4. (Color online) Temperature (K) dependence of lattice and density of modification $Ag_{0.4}Cu_{1.6}Se$: red line — parameters a, b, c (Å) of crystal lattice and density (g/cm) Cu_2Se; blue line — parameters a, b, c (Å) of crystal lattice and density (g/cm) AgCuSe; green line — parameter a (Å) and density (g/cm) of FCC modification $Ag_{0.4}Cu_{1.6}Se$.

Table 4. Thermal expansion factors of $Ag_{0.4}Cu_{1.6}Se$ crystal ($\times 10^6$ K^{-1}).

Modification	T, K	$\alpha_{[100]}$	$\alpha_{[010]}$	$\alpha_{[001]}$	$\bar{\alpha} = \frac{\alpha_{[100]}+\alpha_{[010]}+\alpha_{[001]}}{3}$	$\beta = \alpha_{[100]} + \alpha_{[010]} + \alpha_{[001]}$
Cu_2Se-orthorhombic	295–373	6.043	12.460	26.974	15.159	45.477
	295–423	13.428	15.823	49.589	26.28	78.840
	295–473	1.866	19.809	58.852	32.1	96.519
AgCuSe-orthorhombic	295–373	20.194	47.791	1.610	23.198	69.595
	295–423	35.900	50.348	6.440	30.896	95.686
	295–473	91.311	67.723	−2.800	52.078	156.234
$Ag_{0.4}Cu_{1.6}$Se-FCC	523–573	17.979	17.979	17.979	17.979	53.937

into phases Ag_2Se and AgCuSe. The Cu cations are more than Ag in the $Ag_{0.4}Cu_{1.6}Se$ compound. Therefore, single high-temperature FCC modification below equilibrium temperature ($T_0 = 540$ K) is decomposed into Cu_2Se and AgCuSe.

4. Conclusions

The $Ag_{1.5}Cu_{0.5}Se$ single crystals of high-temperature FCC modification are decomposed into Ag_2Se, AgCuSe and $Ag_{0.4}Cu_{1.6}Se$ when temperature decreases below 488 K and are decomposed into Cu_2Se and AgCuSe when the temperature decreased below $T = 540$ K. Transformations in both compounds are reversible. Temperature dependence of lattice parameters of existing modifications has been used to calculate thermal expansion factors on main crystallographic directions.

References

1. Y. I. Aliyev *et al.*, *J. Ovonic Res.* **14**, 165 (2018).
2. E. B. Asgerov *et al.*, *J. Surf. Invest.: X-ray Synchrotron Neutron Tech.* **12**, 688 (2018).
3. S. H. Jabarov *et al.*, *J. Surf. Invest.: X-ray Synchrotron Neutron Tech.* **9**, 35 (2015).
4. S. H. Jabarov *et al.*, *Materials Science-Poland* **36**, 203 (2018).
5. N. A. Ismayilova and S. H. Jabarov, *Optoelectron. Adv. Mater.-Rapid Commun.* **11**, 353 (2017).
6. Y. I. Aliyev *et al.*, *Mod. Phys. Lett. B* 1850128 (2019).
7. Y. I. Aliyev *et al.*, *Crystallogr. Rep.* **62**, 610 (2017).
8. Y. G. Asadov *et al.*, *Crystallogr. Rep.* **62**, 618 (2017).
9. Y. I. Aliyev *et al.*, *Semiconductors* **51**, 732 (2017).
10. Y. G. Asadov, Y. I. Aliyev and A. G. Babaev, *Phys. Part. Nuclei* **46**, 452 (2015).
11. M. I. Agaev, S. M. Alekperova and M. I. Zargarova, *Rep. AS Azerbaijan SSR* **27**, 15 (1971).
12. Y. G. Asadov, Y. I. Alyev and K. M. Jafarov, *Inorg. Mater.* **44**, 460 (2008).
13. J. W. Earley, *Am. Mineral.* **35**, 337 (1950).
14. A. J. Frueh, G. K. Cramanske and C. H. Knight, *Z. Kristallogr.* **108**, 389 (1957).
15. S. K. Kiazimov, K. M. Jafarov and Y. G. Asadov, *Phase Transit.* **21**, 11 (1990).
16. Y. G. Asadov *et al.*, *Azerbaijan J. Phys.* **XV**, 89 (2009).
17. D. M. Trots *et al.*, *Eur. Phys. J. B* **51**, 507 (2006).
18. L. Bindi and G. Pratesi, *Can. Mineral.* **43**, 1373 (2005).
19. D. M. Trots, Structure and lattice dynamics of copper- and silver-based superionic conducting chalcogenides. Ph.D. thesis, Darmstadt (Darmstadt Technical University, 2007), p. 118.
20. D. M. Trots *et al.*, *J. Phys.: Condens. Matter.* **20**, 455204 (2008).

https://doi.org/10.1142/9789819818433_0023

Chapter 23

THz Cutoff Frequency and Multifunction $Ti_2Ba_2Ca_2Cu_3O_{10}$/GaAs Photonic Bandgap Materials#

Arafa H. Aly* and Fatma A. Sayed

TH-PPM Group, Physics Department,
Faculty of Sciences, Beni-Suef University, Egypt
**arafa.hussien@science.bsu.edu.eg; arafaaly@aucegypt.edu*

By using the transfer matrix method (TMM), we theoretically explore the transmittance properties and cutoff frequency of one-dimensional photonic crystal (1DPCs) within the terahertz frequency region. The present structure consists of high-temperature superconductor and semiconductor layers. The results of the calculations represent the effects of various parameters on the cutoff frequency. We have used the two-fluid model as well as the Drude model to describe the permittivity of superconductor and semiconductor. Further, we consider that the permittivity of both the materials is depending on the hydrostatic pressure. The present results show that with the increasing of different parameters as the operating temperature, the thickness of semiconductor, and the filling factor of semiconductor, then the cutoff frequency shift to lower frequencies regions. By the increasing of superconductor thickness, hydrostatic pressure, doping concentration and filling factor of the superconductor, we found the cutoff frequency shifts to higher frequency regions. These results indicate that cutoff frequency can be modified through these different parameters. Finally, the present design could be useful for many optical systems as the optical filter, reflector and photoelectronic applications in the Terahertz regime.

Keywords: Cutoff frequency; doping concentration; hydrostatic pressure; two-fluid model; Drude model; transfer matrix method; photonic crystal.

*Corresponding author.

#To cite this article, please refer to its earlier version published in the *International Journal of Modern Physics B*, Volume 34, 2050091 (2020), DOI: 10.1142/S0217979220500915.

1. Introduction

Photonic crystals (PCs) are artificial structures for controlling and manipulating the propagations of light. One-dimensional photonic crystals (1DPCs) are the most researched and used due to the low cost and variety of applications.[1–3] In PCs, there is an important and unique property which is called photonic bandgaps (PBGs) which prohibit the propagation of light at certain frequency intervals. The PBGs come from the destructive interference of the PC's constituent materials at the interfaces.[3–7] PBG tuning opens a new perspective in science research and technology applications.[4] To get a tunable PBG, some different parameters must rely on the permittivity of one of the constituent materials.[5] The different parameters, such as the temperature, hydrostatic pressure and applying magnetic or electric fields, etc. can alter the structure of these systems and thus the PC's optical response function.[8] Authors designed PC materials by using various materials such as semiconductors,[9–12] metals,[13] superconductors,[3–6,14,15] liquid crystals,[15] meta materials[16] and graphene.[17]

More attention has been drawn to superconducting PCs because of two primary benefits compared to the conventional PC material.[15] First, reducing damping in nonmagnetic material of electromagnetic waves and taking into account the frequency-dependent dielectric permittivity that is well represented in the context of the two-fluid model.[18] The second is that the superconductor's dielectric function depends primarily on the London penetration depth, which is a function of external temperature and external magnetic function. Semiconductors PCs are also tunable and the permittivity of a semiconductor is a function of carrier concentration.[9] In the past few decades, a significant attention was paid to the incorporation of semiconductor materials in photonic device design. Therefore, the inclusion of the superconductor material with semiconductor material in the same PC has an important impact on the PBG modulation.

In this work, we consider $Ti_2Ba_2Ca_2Cu_3O_{10}$ superconductor and GaAs semiconductor materials to construct the 1DPC. For $Ti_2Ba_2Ca_2Cu_3O_{10}$ superconducting material[19,20] that is usually shortened as (TI-2223) and this material has a high-temperature superconductor with critical temperature $Tc = 125$ K. GaAs semiconductor material[1] is a great candidate for high-performance manufacturing in terahertz applications.

The frequency below which electromagnetic waves propagate by PC is called the cutoff frequency. So far, cutoff frequency has been studied extensively. Recently, the properties of cutoff frequency in two-dimensional superconductor PCs.[18] They reported that cutoff frequency affected by many parameters such as filling factor, the lattice constant change, the superconductor threshold frequency and tuned by the operating temperature. In addition, the cutoff frequency is obtained by different structures as follows: in 1D superconductor metamaterial PC and[16,21] in metamaterials PCs within THz.[16] Researchers recently achieved analysis of cutoff frequency in one-dimensional ternary superconducting PC.[21] This work shows that the cutoff frequency is susceptible to the thickness of superconducting materials, the thickness of the dielectric layer, the operating temperature and the intermediate dielectric refractive index. Cutoff frequency tuning in 1DPCs comprises ($HgBa_2Ca_2Cu_3O_8$) and GaAs.[1] The simulation obviously shows that the cutoff frequency can be manipulated by the constituent materials' thicknesses, pressure and operating temperature.

Motivated by the above work, we describe the transmittance property of a 1D binary PC composed of superconductor and semiconductor in the terahertz frequency region. We used the transfer-matrix method (TMM), two-fluid models and a simple Drude model in this communication theoretically. The THz frequency range is of specific concern as waves can penetrate clothing, paper, wood, plastics and some strong items with these frequencies.

Here, using the well-known superconductor and semiconductor, we investigate the appearance of cutoff frequency within the THz range. Besides, we demonstrate the number of periods, operating temperature, the material thickness effect of the constituents, hydrostatic pressure, doping concentration of semiconductor material and the filling factor of constituent materials on the cutoff frequency. So far, there are no reported papers on the study of the effect of these parameters on ($Ti_2Ba_2Ca_2Cu_3O_{10}$) superconductor material with the semiconductor of the PC.

This research is structured as follows: the fundamental models are discussed in Sec. 2. We present the numerical outcomes described in Sec. 3. Finally, the conclusions of this study are presented in Sec. 4.

2. Basic Equations

We will obtain the transmittance characteristics of the present periodic structure and we use the well-known TMM. As shown in Fig. 1, the structure is conceived as $(AB)^N$, where layer A represents the high Tc superconductor layer; ($Ti_2Ba_2Ca_2Cu_3O_{10}$).[20] Layer B is a semiconductor

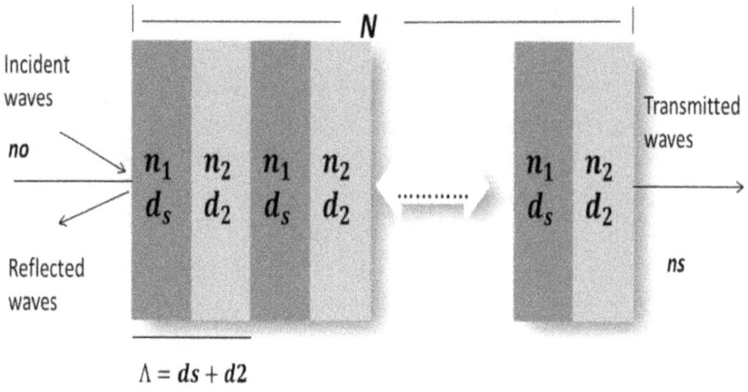

Fig. 1. (Color online) OD PC schematic diagram consisting of a superconductor and semiconductor materials for N-period where the superconductor layer is layer one with thickness d_s and semiconductor layer two with thickness d_2 where the period of the lattice is Λ. The substrate is taken to be glass with a refractive index $ns = 1.52$ and refractive index of incident electromagnetic waves on structure taken to be no = 1.

layer (GaAs).[1,2] This structure is repeated for N-periods and the entire structure is immersed in the air and substratum.

We have used two-fluid model,[3,20] where the dielectric constant of the superconductor layer without losses is given by

$$\epsilon_{(w)} = 1 - \frac{c^2}{w^2 \lambda_L^2}, \tag{1}$$

where c is the light speed in the vacuum, w is the angular frequency and where λ_L is the temperature-dependent penetration depth as described as follows:[1]

$$\lambda_L(T) = \frac{\lambda_O}{\sqrt[3]{\left(1 - \left(\frac{T}{Tc}\right)\right)^q}}, \tag{2}$$

where λ_O is the London-penetration depth at $T = 0$ K, T is the operating temperature, Tc is the critical temperature and q take value equals 3 for a high-temperature superconductor. The refractive index of the superconductor layer is $n_1 = \sqrt{\epsilon_{(w)}}$.

The hydrostatic pressure (P) impact of Tc is accounted for as follows:

$$Tc = x_1 + x_2 P + x_3 P^2, \tag{3}$$

where $x_1 = 125$, $x_2 = 2.009$ and $x_3 = -4.194 \times 10^{-2}$ are simply fitting constants[2] therefore λ_L is also modified due to Tc.

Similarly, the thickness of the superconducting layer (d_s) is changed according to its constant compressibility $K_a = 2.57 \times 10^{-3}$ GPa^{-1}[21] and we suppose here that λ_O will remain constant.

In the regime of terahertz, the dielectric constant of semiconductor layer described by the simple Drude model,[6,7,13] which is given by

$$\in_{(D)}= \epsilon_{(\infty)}\left(1 - \frac{w_p^2}{(w^2 - iw\tau)}\right), \tag{4}$$

where $\epsilon_{(\infty)}$ is dielectric constant at the high frequency, τ is scattering times, and w_p is plasma frequency and $w_p = \sqrt{Ne^2/\in_o m^*}$ which is dependent on the intrinsic carrier density (N), the effective mass of free carriers (m^*), vacuum permittivity ($\in_o$) and charge of carrier (e). The refractive index of the semiconductor layer is $n_2 = \sqrt{\in_{(D)}}$.

To include the variation in the thickness of the GaAs layer and its dielectric constant as a function of pressure and temperature.[22–24]

(a) The thickness of the GaAs layer as a function of pressure is given by

$$d_2(P) = d_2(1 - (f_{11} + 2f_{12})P), \tag{5}$$

where d_2 is the original thickness of the semiconductor layer, $f_{11} = 1.16 \times 10^{-2}$ GPa^{-1} and $f_{12} = -3.7 \times 10^{-3}$ GPa^{-1} are the elastic constants of GaAs.[8]

(b) The dielectric constant under both the pressure and temperature is given by

$$\in_b (P, T) = \left(\in_{(o)} + Ae^{\frac{T}{T_o}}\right) e^{-\beta P}, \tag{6}$$

where $\in_{(o)}= 12.446$, $T_o = 240.7$ K, $A = 0.21125$ and $\beta = 0.0173$ GPa^{-1}.

The frequency-dependent on dielectric constant of the GaAs is given by[2,8]

$$\in_b^* = \in_b (P, T) \times \in_{(D)}. \tag{7}$$

We suppose that w_p weakly dependent on the hydrostatic pressure.

The transmittance spectrum can be calculated by the TMM.[25] According to TMM, the total transfer matrix takes the form.

$$\begin{pmatrix} m_{11} & m_{12} \\ m_{21} & m_{22} \end{pmatrix} = D_A^{-1}(D_1P_1D_1^{-1}D_2P_2D_2^{-1})^N D_A, \tag{8}$$

P_i ($i = 1$, 2 and A) is the propagation matrix in layer i which is expressed as the following equation[6]:

$$P_i = \begin{bmatrix} \exp(jk_id_i) & 0 \\ 0 & \exp(-jk_id_i) \end{bmatrix}, \tag{9}$$

$$k_i = \frac{\omega n_i \cos\theta_i}{c} \quad \text{is the wavenumber,} \tag{10}$$

D_i (for $i = 1, 2$) is the dynamical matrix and given by Eqs. (11) and (12), where

$$D_q = \begin{bmatrix} 1 & 1 \\ n_q\cos\theta_q & -n_q\cos\theta_q \end{bmatrix} \quad \text{for (TE) component,} \tag{11}$$

$$D_q = \begin{bmatrix} \cos\theta_q & \cos\theta_q \\ n_q & -n_q \end{bmatrix} \quad \text{for (TM) component.} \tag{12}$$

The transmittance T is determined by the following relationship, using M_{11} element of matrix from Eq. (8).

$$T = \left|\frac{1}{m_{11}}\right|^2. \tag{13}$$

3. Results and Discussion

In our simulation, we have considered the superconductor and the semiconductor materials to be $Ti_2Ba_2Ca_2Cu_3O_{10}$ (Tc = 125 K, λ = 200 nm) and GaAs ($\epsilon_{(\infty)}$ = 11.7), respectively. In the present calculations, we have considered the normal incidence and we used TE waves. First, we study the dependency of cutoff frequency on the number of periods. Second, we investigate the effect of operating temperature, the thickness of superconductor and semiconductor layers, the hydrostatic pressure and the doping concentration of GaAs. Finally, we study the dependency of the cut-off frequency on the filling factor of superconductor and semiconductor materials.

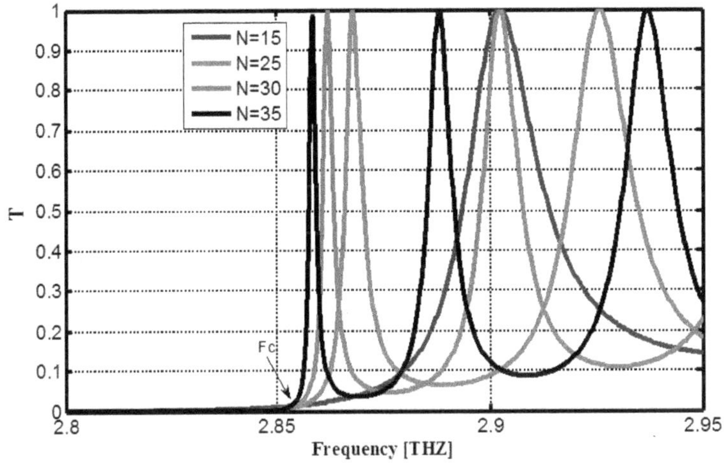

Fig. 2. (Color online) Transmittance spectra at different number of periods with ($Ti_2Ba_2Ca_2Cu_3O_{10}$) layer thickness $d_s = 10$ nm, GaAs layer thickness $d_2 = 5000$ nm, $T = 4$ K and $P = 0$ GPa.

Figure 2 demonstrates the transmittance spectrum at different number of periods ($N = 15$, 25, 30 and 35), where the operating temperature of superconductor $T = 4$ K, thickness of superconductor and semiconductor ($d_s = 10$ nm and $d_2 = 5000$ nm). In the absence of hydrostatic pressure in the region of terahertz frequency, the figure illustrates the appearance of the cutoff frequency (Fc) at THz frequencies. In addition to, the figure shows two points, the first point as the number of periods increases, the value of (Fc) slightly decreases, the second point the spectrum of transmittance becomes sharper and steeper as the number of periods increases. These results demonstrate that when their frequency is lower than the cutoff frequency, the propagation of the electromagnetic waves can be prohibited. In other words, the structure can act as a good reflector. Therefore, we will set $N = 25$ in the following calculations.

3.1. *The effect of operating temperature on the superconductor layer*

The transmittance spectra are shown in Fig. 3 where the operating temperature varies ($T = 4$ upto 90). The thickness of superconductor and semiconductor materials was assumed to be $d_s = 10$ nm and $d_2 = 5000$ nm, respectively. Here, we also considered the absence of pressure.

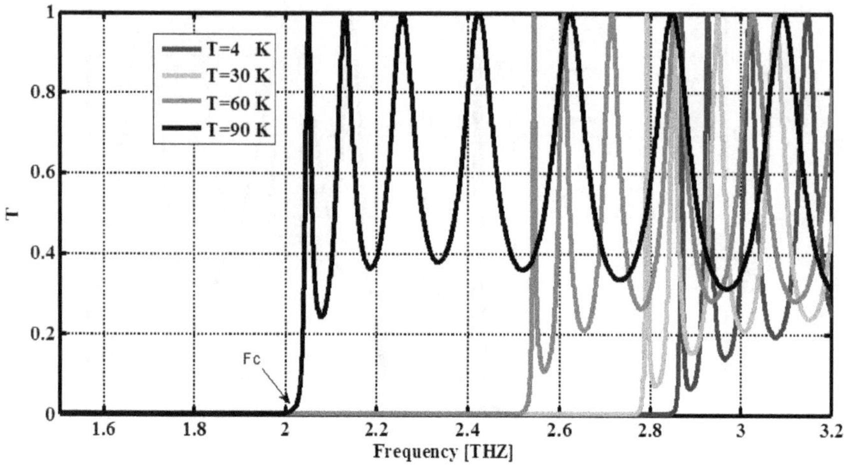

Fig. 3. (Color online) Transmittance spectra at different temperatures with ($Ti_2Ba_2Ca_2Cu_3O_{10}$) layer thickness $d_s = 10$ nm, GaAs layer thickness $d_2 = 5000$ nm, $P = 0$ GPa and number of periods $N = 25$.

Table 1.

Operating temperature (K)	Position of cutoff frequency (THZ)
$T = 4$ K	Fc $\approx$ 2.87 THZ
$T = 30$ K	Fc $\approx$ 2.78 THZ
$T = 60$ K	Fc $\approx$ 2.54 THZ
$T = 90$ K	Fc $\approx$ 2.04 THZ

Based on the theoretical model, there is a clear dependence of the superconducting permittivity on the operating temperature. There is a significant variation in the transmittance spectra owing to variations in operating temperature as shown in Fig. 3. The results showed that the increase in the operating temperature of the superconductor material leads to a change in the amplitude of the oscillation. Besides, the cutoff frequency shifts the low-frequency regions. We note that, at $T \leq 4$ K, the cutoff frequency is slightly affected by the change in operating temperature, but when operating temperature approaches critical temperature, the cutoff frequency values are sharply decreased as shown in Table 1. The major reduction in the cutoff frequency values may be due to the minimization in the number of super electrons over the normal number. This result indicates that tuning the structure can be achieved by changing the operating temperature. This is a fantastic benefit wherein superconductor electronics and temperature-dependent PCs could be useful. The following calculation uses the superconductor material at high-temperature $T = 90$ K near to critical temperature, due to superconductor material more sensitive at a temperature near to its critical temperature.

3.2. *The effect of superconductor layer thickness*

In the next step, we have studied the transmittance spectra behavior at different thicknesses of the superconductor ($d_s = 10, 30, 60, 90$ and 120 nm) in Fig. 4 and the thickness of semiconductor was fixed at $d_2 = 5000$ nm. We have used the temperature was set at $T = 90$ K, $N = 25$ and $P = 0$ GPa. It's obvious from Fig. 4 that the cutoff frequency is shifted to higher frequency values, as shown in Table 2 with increasing the thickness of the superconductor layer. However, the increase in the superconductor material's thickness results in a significant shift in the amplitude of the oscillation as shown in Fig. 4 so the cutoff frequency shifts to a higher frequency. We note that the cutoff frequency has different responses compared to the numerical results investigated in Fig. 3. Therefore, the

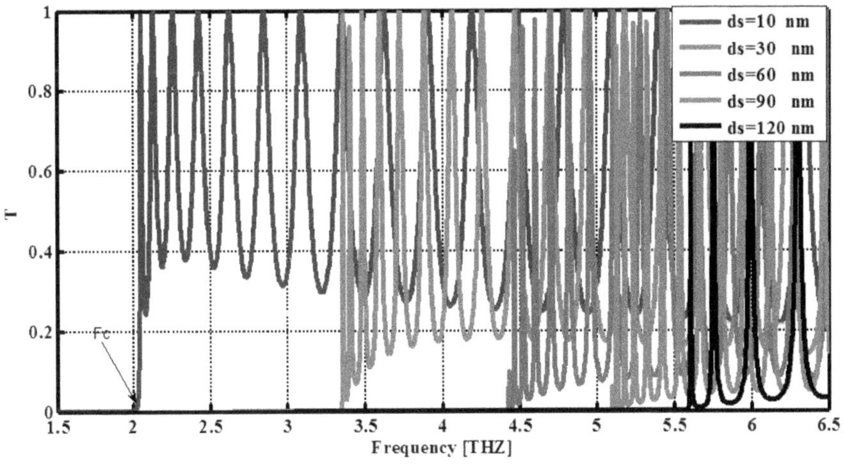

Fig. 4. (Color online) Transmittance spectra of the structure with ($Ti_2Ba_2Ca_2Cu_3O_{10}$) layer thickness d_s varied to be 10, 30, 60, 90 and 120 nm, while semiconductor thickness d_2 is fixed at 5000 nm. The temperature, pressure and the number of periods are 90 K, 0 GPa and 25, respectively.

Table 2.

Thickness of superconductor layer (nm)	Position of cutoff frequency (THZ)
$d_s = 10$ nm	Fc $\approx$ 2.03 THZ
$d_s = 30$ nm	Fc $\approx$ 3.35 THZ
$d_s = 60$ nm	Fc $\approx$ 4.46 THZ
$d_s = 90$ nm	Fc $\approx$ 5.13 THZ
$d_s = 120$ nm	Fc $\approx$ 5.60 THZ

cutoff frequency decreases with the increased operating temperature of the superconductor and shifts to lower frequency regions but increases with increasing thickness of the superconductor layer. If we compared this structure with another structure as in Ref. 2, then we note that the thickness of the superconductor was used in our calculations is very small ($d_s = 10$ nm) compared to the thickness of superconductor ($d_s = 5000$ nm) used in Ref. 2. Maybe the present design is useful in applications compared with another.

3.3. *The effect of semiconductor layer thickness*

In Fig. 5, we studied the transmittance spectrum behavior at different thicknesses of the semiconductor layer ($d_2 = 1000, 2000, 3000, 4000$ and 5000 nm) and thickness of superconductor layer was fixed at $d_s = 10$ nm, we note that the cutoff frequency shifts to lower values. This means that with increasing thickness of the semiconductor layer, the cutoff frequency decreases, as shown in Table 3, unlike the thickness of the superconductor. However, the increase in semiconductor material thickness results in a significant increase in the amplitude of the oscillation. The transmission spectrum becomes sharper and steeper as the thickness of the semiconductor increases. We can conclude from these results that this structure can operate as a high-pass filter, as shown in Figs. 4 and 5. It can be seen that the cutoff frequency is more influenced by the thickness of the superconductor layer than the thickness of the semiconductor layer.

3.4. *The effect of the hydrostatic pressure*

We discuss the effect of hydrostatic pressure on the cutoff frequency in the proposed structure. Figure 6 displays the transmittance spectrum when the hydrostatic pressure increases ($P = 0$, 10, 20, 30 and 40 GPa)

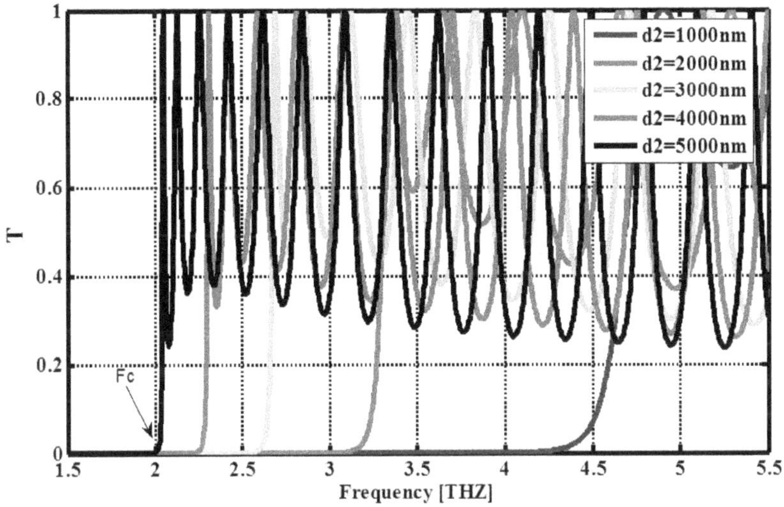

Fig. 5. (Color online) Transmittance spectra of the structure with GaAs layer thickness d_2 equal to 1000, 2000, 3000, 4000 and 5000 nm, while the superconductor thickness d_s fixed at 10 nm. The temperature, pressure and the number of periods are 90 K, 0 GPa and 25, respectively.

Table 3.

The thickness of semiconductor layer (nm)	Position of cutoff frequency (THZ)
$d_2 = 1000$ nm	Fc $\approx$ 4.6 THZ
$d_2 = 2000$ nm	Fc $\approx$ 3.27 THZ
$d_2 = 3000$ nm	Fc $\approx$ 2.66 THZ
$d_2 = 4000$ nm	Fc $\approx$ 2.29 THZ
$d_2 = 5000$ nm	Fc $\approx$ 2.03 THZ

where the operating temperature, the thickness of superconductor and semiconductor, and the number of periods are 90 K, 10, 5000 nm and 25, respectively. In Fig. 6, the value of cutoff frequency shifts toward with increasing hydrostatic pressure; this means that cutoff frequency increases with increase in hydrostatic pressure as shown in Table 4. However, the increase in hydrostatic pressure leads to an increase in the amplitude of oscillation of the transmission. In addition, hydrostatic pressure effect on the thickness of semiconductor as in Eq. (5) and with increase hydrostatic pressure, the thickness of the semiconductor layer decreases. From Table 4,

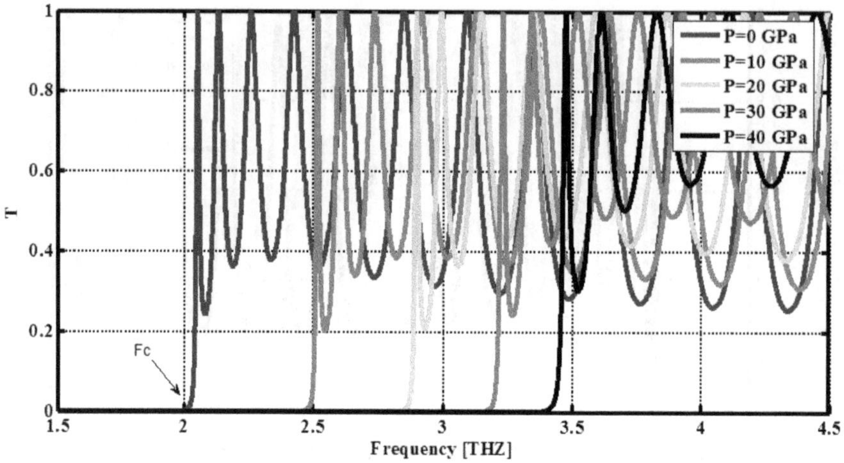

Fig. 6. (Color online) The cutoff frequency for different pressures with $d_s = 10$ nm and $d_2 = 5000$ nm. The number of periods and temperature are 25 and 90 K, respectively.

Table 4.

Hydrostatic pressure (GPa)	Position of cutoff frequency (THZ)
$P = 0$ GPa	Fc $\approx$ 2.036 THZ
$P = 10$ GPa	Fc $\approx$ 2.50 THZ
$P = 20$ GPa	Fc $\approx$ 2.88 THZ
$P = 30$ GPa	Fc $\approx$ 3.21 THZ
$P = 40$ GPa	Fc $\approx$ 3.45 THZ

we note that the cutoff frequency decreases with increasing hydrostatic pressure value, when P increases from 0 GPa to 10 GPa, the cutoff frequency increases by 0.46 THz, but when P increases from 30 GPa to 40 GPa, the cutoff frequency increases by 0.24 THz.

3.5. *The effect of the doping concentration of the semiconductor material*

The dependence of the transmittance spectrum on the doping concentration of the semiconductor material is illustrated in Fig. 7, which the superconductor thickness is 10 nm and semiconductor thickness is 5000 nm with

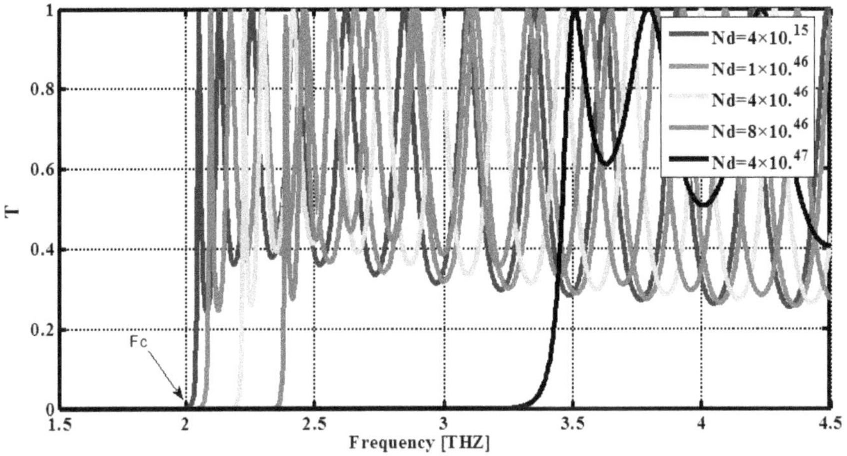

Fig. 7. (Color online) The cutoff frequency at different values of the doping concentration of the semiconductor layer with $d_s = 10$ nm and $d2 = 5000$ nm. The temperature, pressure and the number of periods are 90 K, 0 GPa and 25, respectively.

Table 5.

Doping concentration	Position of cutoff frequency (THZ)
$N_d = 4 \times 10^{15}$	Fc $\approx$ 2.035 THZ
$N_d = 1 \times 10^{46}$	Fc $\approx$ 2.08 THZ
$N_d = 4 \times 10^{46}$	Fc $\approx$ 2.21 THZ
$N_d = 8 \times 10^{46}$	Fc $\approx$ 2.37 THZ
$N_d = 4 \times 10^{47}$	Fc $\approx$ 3.41 THZ

$T = 90$ K and $N = 25$. We note that cutoff frequency shifted toward higher frequency regions with increasing doping concentration of semiconductor layer. Although the shift to higher frequency is not regular as shown in Table 5, we found when N_d increased from $(4 \times 10^{15}$ to $1 \times 10^{46})$, the cutoff frequency increased by the value (0.04 THz), but when N_d increased from $(8 \times 10^{46}$ to $4 \times 10^{47})$, the cutoff frequency increased by the value (1.04 THz). This result indicates that the present design did not get affected by the increase of semiconductors doping concentration from the initial doping concentration which made it (4×10^{15}) until to reach (1×10^{46}), also at this value, the cutoff frequency will start to increase.

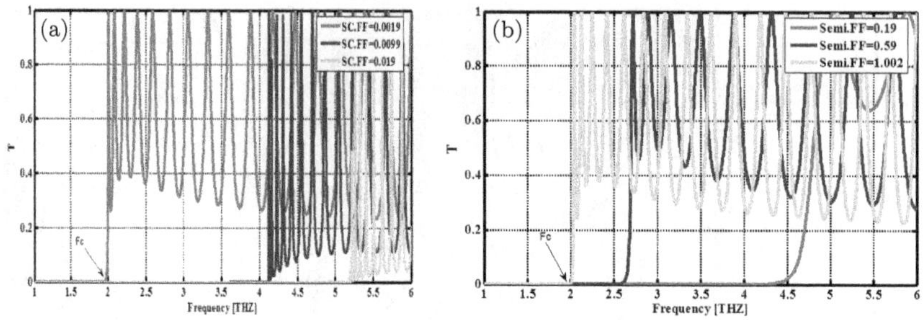

Fig. 8. (Color online) Calculated cutoff frequency of two SC/Semi-PCs with $\Lambda = 5010$ nm, $T = 90$ K and different only in the filling factors of the constituent materials. (a) At different filling factors of the superconductor and (b) at different filling factors of the semiconductor.

3.6. *The effect of filling factors*

Finally, we study the effect of varying the filling factor of a superconductor (sc.ff) and semiconductor (semi.ff) on the transmittance spectra as shown in Figs. 8(a) and 8(b), where the filling factor (f.f) of a material in a PC can be given by this relation $f = d/\Lambda$[9]; where d is the material thickness and Λ is the lattice constant. As shown in Fig. 8(a), with increasing the value of superconductor filling factor, the cutoff frequency is shifted toward high-frequency regions, that is the cutoff frequency is directly proportional to the superconductor filling factor. Besides, with the increase in the superconductor filling factor, the amplitude of oscillations is increased. In Fig. 8(b), with an increase in the semiconductor filling factor, cutoff frequency goes downward to lower frequency regions, this means the cutoff frequency is inversely proportional to semiconductor filling factor unlike the behavior in Fig. 8(a). We note that in Fig. 8(b), the amplitude of oscillations begins to grow with increasing the semiconductor filling factor.

4. Conclusion

We have investigated the transmittance properties of 1DPCs consisting of a superconductor ($Ti_2Ba_2Ca_2Cu_3O_{10}$) and semiconductor (GaAs) material in the terahertz frequency. Our technique of calculation is based on two-fluid models, the simple Drude model, and the TMM. Numerical results show the cutoff frequency under which electromagnetic waves cannot propagate within PCs. Also, the cutoff frequency can be tuned by many parameters. As the operating temperature or thickness of the semiconductor or filling factor of semiconductor is increased, the value of cutoff frequency is shifted to downward. Whereas the thickness of superconductor or hydrostatic pressure or doping concentration is increased, the value of cutoff frequency is shifted to toward. We noted that the transmittance spectrum became sharper and steeper, as the number of periods increased. Therefore, the superconductor semiconductor PC can work as a very useful frequency selector as well as a bandpass filter in the THz region, whereas it acts as a good reflector for frequencies below the cutoff frequency.

References

1. F. S. Chaves and H. V. Posada, *Physica C: Supercond. Appl.* **18**, 30462 (2019).
2. A. Y. Herrera, J. M. Calero and N. P. Montenegro, *J. Appl. Phys.* **123**, 033101 (2018).
3. I. S. Amiri *et al.*, *Microw. Opt. Technol. Lett.* **61**, 847 (2019).
4. A. H. Aly, A. A. Ameen and D. Vigneswaran, *J. Supercond. Nov. Magn.* **32**, 191 (2019).
5. N. R. Ramanujam *et al.*, *Physica B* **572**, 42 (2019).
6. A. H. Aly and D. Mohamed, *J. Supercond. Nov. Magn.* **32**, 1897 (2019).
7. V. S. Revathy *et al.*, *Optik* **176**, 78 (2019).
8. H. A. Elsayed and A. H. Aly, *Int. J. Mod. Phys. B* **32**, 1850056 (2017).
9. A. H. Aly, D. Mohamed and H. A. Elsayed, *Lett. Appl. NanobioSci.* **3**, 2284 (2014).
10. C. Liu and C. Jang, Analysis of Multiple Resonant Tunneling in a Strongly Dispersive Semiconductor Photonic Crystal, Institute of Electro-Optical Science and Technology (2013).
11. L. E. González and N. P. Montenegro, *Physica E* **44**, 773 (2012).
12. A. H. Aly, M. Ismaeel and E. A. Rahman, *Opt. Photonics J.* **2**, 105 (2012).
13. F. S. Chavesa and H. V. Posadaa, *Int. J. Light Electron Opt.* **181**, 493 (2019).
14. F. S. Chaves and H. V. Posada, *Int. J. Light Electron Opt.* **181**, 416 (2019).
15. P. Singh *et al.*, *Indian Acad. Sci.* **50**, 93 (2019).
16. A. H. Aly, W. Sabra and H. A. Elsayed, *Int. J. Mod. Phys. B* **31**, 1750123 (2017).
17. X. Xie *et al.*, *J. Quant. Spectrosc. Radiat. Transf.* **18**, 30909 (2019).
18. A. H. Aly, H. A. Elsayed and S. A. El-Naggar, *J. Mod. Opt.* **61**, 1064 (2014).
19. W. Buckel and R. Kleiner, *Superconductivity: Fundamentals and Applications*, 2nd edn. (Wiley-VCH Verlag GmbH & Co. KGaA, Weinheim, 2004).
20. K. Kawano *et al.*, *J. Soc. Inorg. Mater.* **7**, 117 (2000).
21. N. M. Dõsouza and V. Mathew, *Physica C: Supercond. Appl.* **17**, 30137 (2017).
22. F. S. Chavesa and H. V. Posada, *Int. J. Light Electron Opt.* **170**, 384 (2018).
23. D. M. Ren, Y. Z. Wei and S. Lei, *Optoelectron. Lett.* **5**, 3 (2012).
24. G. A. Samara, *Phys. Rev. B* **27**, 6 (1983).
25. A. H. Aly *et al.*, *J. Supercond. Nov. Magn.* **31**, 1557 (2018).

https://doi.org/10.1142/9789819818433_0024

Chapter 24

Spatially Resolved Characterization of Optical and Recombination Losses for Different Industrial Silicon Solar Cell Architectures#

E. J. Schneller[1,2,3], M. J. Hossain[2,4], R. Frota[2,4], N. Iqbal[1,2], D. Colvin[1,2,3], A. J. Curran[5,6], M. Wang[5,7], J. L. Braid[5,6], L. S. Bruckman[5,6], R. H. French[5,6,7], B. D. Huey[8], J.-N. Jaubert[9] and K. O. Davis[1,2,3,4,*]

[1]*Department of Materials Science and Engineering, University of Central Florida, Orlando, FL, USA*
[2]*Resilient Intelligent Sustainable Energy Systems Faculty Cluster, University of Central Florida, Orlando, FL, USA*
[3]*Florida Solar Energy Center, University of Central Florida, Cocoa, FL, USA*
[4]*CREOL, The College of Optics and Photonics, University of Central Florida, Orlando, FL, USA*
[5]*SDLE Research Center, Case Western Reserve University, Cleveland, OH, USA*
[6]*Department of Materials Science and Engineering, Case Western Reserve University, Cleveland, OH, USA*
[7]*Department of Macromolecular Science and Engineering, Case Western, Reserve University, Cleveland, OH, USA*
[8]*Department of Materials Science and Engineering, University of Connecticut, Storrs, CT, USA*
[9]*Canadian Solar, Inc., Guelph, Ontario, Canada*
**kristopher.davis@ucf.edu*

In this work, spatially resolved characterization methods are used to identify loss mechanisms for common *p*-type silicon solar cell architectures, including multicrystalline aluminum back surface field (Al-BSF), monocrystalline Al-BSF, monocrystalline passivated emitter and rear cells (PERC), and bifacial monocrystalline PERC. The characterization methods used in this work include

*Corresponding author.

#To cite this article, please refer to its earlier version published in the *International Journal of Modern Physics B*, Volume 34, 2050204 (2020), DOI: 10.1142/S0217979220502045.

suns-V_{OC}, photoluminescence imaging, and spatially resolved external quantum efficiency and reflectance measurements. The optical and recombination losses are driven by the material properties, cell processing conditions, and device architecture. These losses are quantified and categorized in terms of underlying mechanisms (e.g., front reflectance, escape reflectance, front recombination, and parasitic optical absorption and recombination in the bulk and rear). The ability to create images of these loss parameters can be used to gain more insight into the materials and manufacturing processes used to produce solar cells, and examples are given in this work to illustrate how these images can help reveal the origin of defects.

Keywords: Photovoltaics; silicon; solar cells; metrology; characterization; loss analysis.

1. Introduction

Crystalline silicon (c-Si) solar cells based on *p*-type silicon wafers continue to represent the overwhelming majority of photovoltaic (PV) products on the market today. The transition from aluminum back surface field (Al-BSF) cells to passivated emitter and rear cells (PERC) has pushed *p*-type cell efficiencies higher than many expected possible.[1–6] PERC efficiency gains are driven by the application of a passivating dielectric stack at the rear surface with local Al-BSF contacts. This dielectric stack reduces the optical and recombination losses and allows cells to have new functionalities, like bifaciality (i.e., the ability to collect light from both the front and rear surface of the cell and module[7,8]).

The transition from Al-BSF to PERC was made possible because the benefits of increased efficiency outweighed the added manufacturing cost associated with additional process steps (e.g., aluminum oxide and silicon nitride rear deposition,[9–18] laser ablation for local contact opening[19–21]). Understanding the loss mechanisms that limit performance is always important in PV manufacturing, but especially so when new materials, processes, or cell designs are being developed and eventually implemented into production. Common metrology methods like illuminated current-voltage (I-V) measurements, *suns*-V_{OC}, external quantum efficiency (EQE), and reflectance (R) are routinely performed on cells to ensure the cells perform to the level expected and diagnose cases where the performance comes up short. Spatially resolved characterization methods can provide much more insight into losses and their root cause. Methods like dark lock-in thermography (DLIT), photoluminescence (PL) imaging, and spatially resolved EQE and R allow scientists and engineers to create images of actual device parameters that can then be used to identify and categorize defects

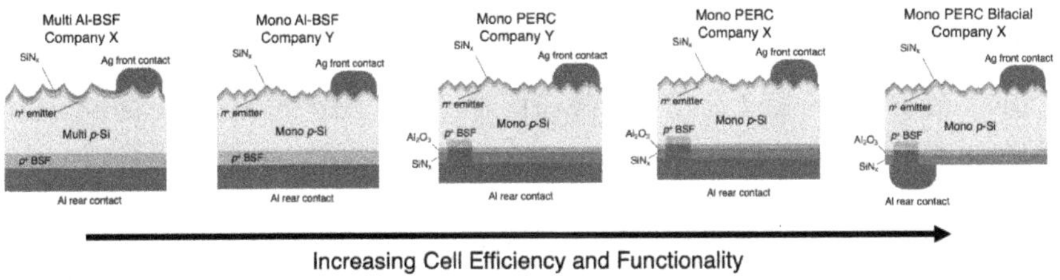

Fig. 1. (Color online) Five groups of industrial c-Si solar cells with varied manufacturing process complexity and cell efficiency.

in PV cells.[22–56] Even when there are no critical defects present, quantifying the spatial homogeneity of device parameters can assist in prioritizing what materials and processes should be improved upon and this data can be used for data mining in production.[57,58]

In order to evaluate the spatial variation of optical and recombination losses in common cell architectures, five groups of industrial solar cells were obtained from various manufacturers. The five groups consist of multicrystalline Al-BSF, monocrystalline Al-BSF, two manufacturers of monocrystalline PERC, and bifacial monocrystalline PERC. In addition to cell-level *suns*-V_{OC} measurements, PL imaging and spatially resolved EQE and R measurements were performed on these cells to demonstrate how these methods can be used in industrial R&D and production.

2. Experimental Details

For this work, a subset of six cells was investigated for each of the five cell groups. The cells were acquired from two independent manufacturers, which we will refer to as Company X and Company Y throughout this work. Company X provided a set of multicrystalline Al-BSF, monocrystalline PERC, and bifacial monocrystalline PERC. Company Y provide a set of monocrystalline Al-BSF and monocrystalline PERC. All cells from Company X used a five busbar cell metallization scheme, while all cells from Company Y used a four busbar cell metallization scheme.

Characterization of these cells were carried out using a Sinton Instruments *suns*-V_{OC} system, a Tau Science FlashQE system, and a BT Imaging LIS-R1 PL imaging system. The Sinton Instruments *suns*-V_{OC} system was used to measure the V_{OC} at one sun, the dark saturation current density (J_{01}), and the effective carrier lifetime of the cells (τ_{eff}) at the pseudo-maximum power point.[59,60]

To obtain spectrally and spatially resolved data, the FlashQE system was used.[39,40,61,62] This system utilizes an array of LEDs to obtain a full EQE and R curve in 1 S. In this work, the spot size for this measurement was approximately 5 mm in diameter. To ensure consistency with respect to the gridline shading for each measurement, the step size was set equal to the pitch between gridlines. The system then scanned the full 156 cm × 156 cm cell with approximately 100 × 100 resolution, resulting in approximately 10,000 spectral response curves per each cell. At each point, the EQE spectra is multiplied by the available photon flux at AM1.5G and the product integrating over the solar spectrum to obtain the short-circuit current density (J_{SC}). To obtain current loss terms at each point, a similar process is applied to the R spectra and the (1-EQE-R) spectra by integrating over a confined range of wavelengths based on the loss mechanism. A detailed discussion of how this analysis is carried out using this FlashQE system can be found in previous work.[62]

Finally, the BT Imaging system was used to obtain PL images. For each cell at least one image was collected for the cells using a 808 nm laser, at an intensity equivalent to one sun, with the cell in the open circuit condition. To approximate V_{OC} images, the natural logarithm of the luminescence intensity image was used and then scaled linearly based on the global cell V_{OC}.

3. Results and Discussion

3.1. *Global cell performance*

As expected, the global cell performance of the PERC devices outperformed the traditional Al-BSF devices. The distribution of V_{OC} and J_{SC} for each of the cell groups is shown in Fig. 2(a). The transition to PERC corresponds to an increase in V_{OC} in the range of 20 mV–25 mV. Along with the boost in V_{OC}, there is also a substantial increase in the J_{SC} of these devices of just over 1 mA/cm2 compared to their monocrystalline Al-BSF counterparts. To compare the difference between monocrystalline and multicrystalline wafers, we can look at the two Al-BSF cell groups. Although there is a slight improvement in the V_{OC} for the monocrystalline Al-BSF group, this difference is smaller than what might be expected. This is likely due to the differences in manufacturing processes between the two groups as we will discuss in the following section. There is, however, a substantial increase in the J_{SC} for the monocrystalline Al-BSF group. The random upright pyramids formed by the alkaline texturing process for {100} oriented

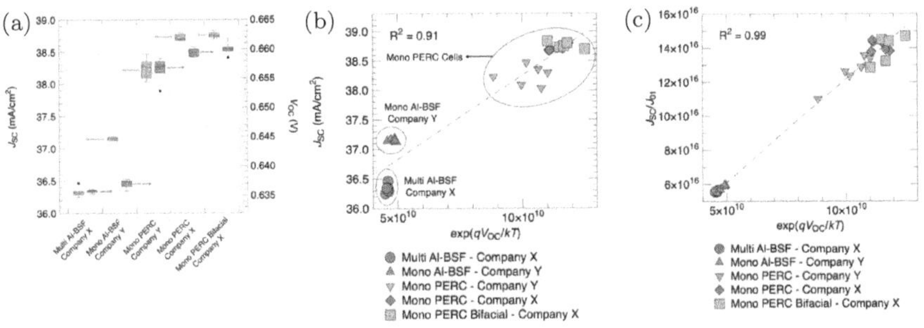

Fig. 2. (Color online) (a) Box-plot for each of the five cell groups showing the average values and variance in short-circuit current and open circuit voltage. Correlation plots comparing the exponential of open circuit voltage with (b) the short-circuit current and (c) a normalized short-circuit current based on the dark saturation current density J_{01}.

monocrystalline wafers[63–65] is more effective at reducing front reflectance than the isotropic texturing process for multicrystalline wafers.[66] This ultimately allows for an increase in light generated carriers within the device. To compare the various groups, it is instructive to consider the standard one-diode model of a solar cell,

$$J = J_G - J_{01}[e^{\frac{q(V+JR_S)}{kT}} - 1] - \frac{V + JR_S}{R_{\text{SH}}}, \tag{1}$$

where J is the device current density, V is the device voltage, J_G is the light generated current density, J_{01} is the dark saturation current density, and R_S and R_{SH} are the series and shunt resistance of the device, respectively. Often a reasonable approximation of J_G is the short-circuit current of the device J_{SC}. If we then consider the open circuit condition where $J = 0$, and a device with a sufficiently high shunt resistance, we arrive at the following expression:

$$\frac{J_{\text{SC}}}{J_{01}} \approx e^{\left(\frac{qV_{\text{OC}}}{kT}\right)}. \tag{2}$$

In Fig. 2(b), we plot the J_{SC} as a function of the exponential of V_{OC} for the five cell groups. Although there is a fairly strong relationship here, it appears as though the transition from multicrystalline to monocrystalline wafers for the Al-BSF groups leads to substantial deviations. If we then normalize the J_{SC} values based on the measured J_{01} values, the relationship is much stronger which supports the equation above.

A closer look at the EQE and R for one cell within each group allows for an analysis of the spectrally dependent behavior that is responsible

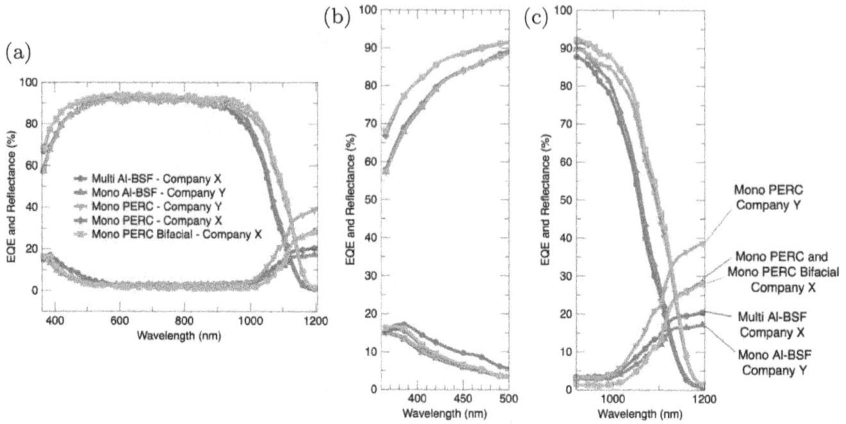

Fig. 3. (Color online) (a) External quantum efficiency and reflectance data for a representative cell in each of the five groups. The short wavelength regime is highlighted in (b) and the long wavelength regime is highlighted in (c).

for variations in J_{SC}. These plots are shown in Fig. 3. The EQE in the short wavelength regime, highlighted in Fig. 3(b), is generally indicative of front surface and emitter effects, while the long wavelength regime, highlighted in Fig. 4(c), is indicative of the base of the device including bulk recombination, rear surface recombination, and parasitic absorption at the rear (i.e., within the rear contact layers). The short wavelength EQE appears to consist of two distinct groups. The cells with the lower blue response correspond to both groups from Company Y and the multicrystalline Al-BSF group. The multicrystalline Al-BSF group also shows a marked increase in reflectance, largely accounting for this reduction in EQE. The lower blue response in the other two groups from Company Y indicates either a higher dopant concentration within the emitter or lower quality front surface passivation.

The long wavelength EQE is also clearly split into two groups with all three PERC groups showing a better response than the two Al-BSF groups. The long wavelength gains observed in the PERC groups are due to a reduction in the overall rear surface recombination of the device along with an increase in the internal rear reflectance.[67–72] We can observe changes in the internal rear reflectance by measuring the reflectance near the band edge, which is commonly referred to as escape reflectance. This corresponds with light that enters the cell, travels through the width of the device, reflects off of the rear surface of the device, travels back through the width of the device and finally exits the cell on the front side. For the monocrystalline PERC cells, the combination of the pyramid structure on the front that alters the propagation angle of light entering the c-Si absorber and the large difference in the refractive index between Si and the underlying dielectrics (i.e., Al_2O_3, SiN_x) result in frustrated total internal reflection. This is why the PERC groups all feature a higher reflectance at wavelengths approaching the band gap. Another competing optical phenomenon that influences the escape reflectance is the front internal reflectance. This refers to the case where light is internally reflected at the rear and is defined as the internal reflectance at the front surface of the device for the light propagating back up to the front. A combination of high rear and front internal reflectance is what leads to effective light trapping. Increasing the internal front reflectance can be accomplished in PERC cells by enabling light scattering at the rear.[73] As the front internal reflectance increases, the magnitude of escape reflectance would reduce, and is the likely cause for the difference in escape reflectance between the two PERC manufacturers.

3.2. *Optical and recombination losses by mechanism*

The global cell performance can be understood by examining the underlying processes impacting device behavior. It is generally instructive to break down loss mechanisms into three distinct groups: optical, recombination, and resistive. Optical losses generally impact the current of the device. Recombination losses are closely tied to the open circuit voltage of the device, but can also influence other regions of the J-V curve. Resistive effects are dominated by the physical properties of the metallization and to a lesser extent, the emitter sheet resistance. These resistive losses mainly influence the fill factor of the device, which was not investigated in this work.

The open circuit voltage is useful for investigating recombination-related losses because V_{OC} is virtually unaffected by both optical and resistive effects, making *suns*-V_{OC} such a powerful metrology technique. Fig. 4(a) shows two metrics derived from the *suns*-V_{OC} measurement: τ_{eff} and J_{01}, which are closely related recombination parameters that show an inverse relationship with each other, as expected. As we can see, the improvement in V_{OC} of the PERC cells are directly linked with both of these metrics. The rear side passivating dielectric stack reduces the recombination rate at this interface, reducing J_{01} and increasing τ_{eff}. For all three PERC groups, an increase in the variance of τ_{eff} among the samples was also observed. Likely,

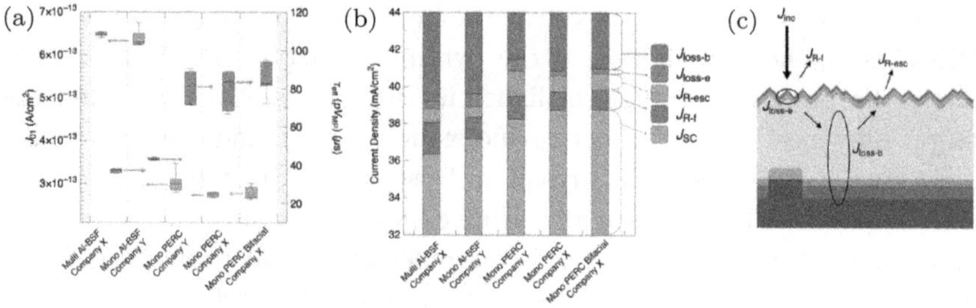

Fig. 4. (Color online) (a) Effective carrier lifetime (τ_{eff}) at the pseudo-maximum power voltage and dark saturation current density (J_{01}) as determined from *suns*-V_{OC} measurements. (b) Stacked bar plots of short-circuit current density (J_{SC}) along with four distinct current loss mechanisms generated from the average of spatially resolved EQE and R measurements. (c) A cell cross-section illustrating each of the current loss mechanisms: emitter loss ($J_{\text{loss}-e}$), base loss ($J_{\text{loss}-b}$), front surface reflectance (J_{R-f}), and escape reflectance ($J_{R-\text{esc}}$).

now that the rear surface recombination rate is no longer the limiting factor, variations in the other contributing factors for recombination, including bulk defects in the wafer, emitter dopant concentration, or the front surface passivation have become more pronounced.

A difference in the recombination properties of the multicrystalline and monocrystalline Al-BSF groups was also observed. Grain boundaries in multicrystalline wafers are known to have higher rates of recombination than the bulk c-Si. Therefore, we would expect a higher τ_{eff} for the monocrystalline group, and we do indeed observe this. The difference, however, is not particularly large. The quality of multicrystalline wafers has been improving at a remarkable pace in recent years, which has to some extent minimized the advantage of monocrystalline wafers. There are also, however, contributions from other recombination losses that could be different between the two manufacturers. Most notably is the emitter recombination which we observe to be higher in the samples from Company Y. To confirm this, a transmission line measurement (TLM) was performed on a laser-scribed strip from one sample in each group to extract the emitter sheet resistance. The results are shown in Table 1 and confirm a higher emitter dopant concentration (i.e., lower sheet resistance) in all of the samples from Company Y, including the monocrystalline Al-BSF group.

To evaluate spectral response data, it is instructive to consider the maximum possible current and break it down into various categories as is done in Fig. 4(b). For this work, the available photons in the

Table 1. Sheet resistance measurements extracted from TLM data for one representative sample in each of the five groups.

Manufacturer	Sample Type	Sheet Resistance (Ω/#sq.)
Company X	Multicrystalline Al-BSF	110.03
	Monocrystalline PERC	107.24
	Monocrystalline PERC Bifacial	112.66
Company Y	Monocrystalline Al-BSF	69.3
	Monocrystalline PERC	75.7

AM1.5G spectrum from 365 nm to 1195 nm was considered the upper limit for maximum possible current. Using a methodology described in previous work, current losses were broken down into four distinct categories: emitter loss ($J_{\text{loss}-e}$), base loss ($J_{\text{loss}-b}$), front surface reflectance (J_{R-f}), and escape reflectance ($J_{R-\text{esc}}$).[62] These losses are depicted in Fig. 4(c). The emitter loss directly correlates with the TLM data, where the devices with the lower sheet resistance also have a relatively higher $J_{\text{loss}-e}$. The higher dopant concentration allowing for the lower sheet resistance results in a current loss penalty of about 0.3 mA/cm2. The J_{R-f} data indicate that the surface texture on the multicrystalline wafers has a penalty of about 0.6 mA/cm2 as compared to the monocrystalline wafers. When considering the base loss, which includes parasitic absorption in the bulk of the device and at the rear surface along with recombination losses in the bulk and at the rear, it is critical to compare it in relation to the other current loss mechanism. For example, the cell with the lowest base loss is the monocrystalline PERC from Company Y. This group, however, has the lowest overall J_{SC} for all of the PERC groups. Each of the other three loss mechanisms are higher for this group as compared to the other PERC groups, meaning that there are less overall photons available within the base of the device. For these devices, it is useful to consider the effective diffusion length in the base of the device (L_{eff}). The longer the L_{eff}, the more likely a carrier will make it to the junction and contribute to the device current. For this metric, all of the PERC groups were very similar, except for a few cells with some spatial nonuniformities in the group from Company Y that will be discussed in the next section.

Many of the loss metrics discussed here are related to each other. To examine these relationships, correlation plots were generated and are shown in Fig. 5. Key metrics from the *suns*-V_{OC} measurements, including τ_{eff} and J_{01}, are compared to key metrics from the spatially resolved

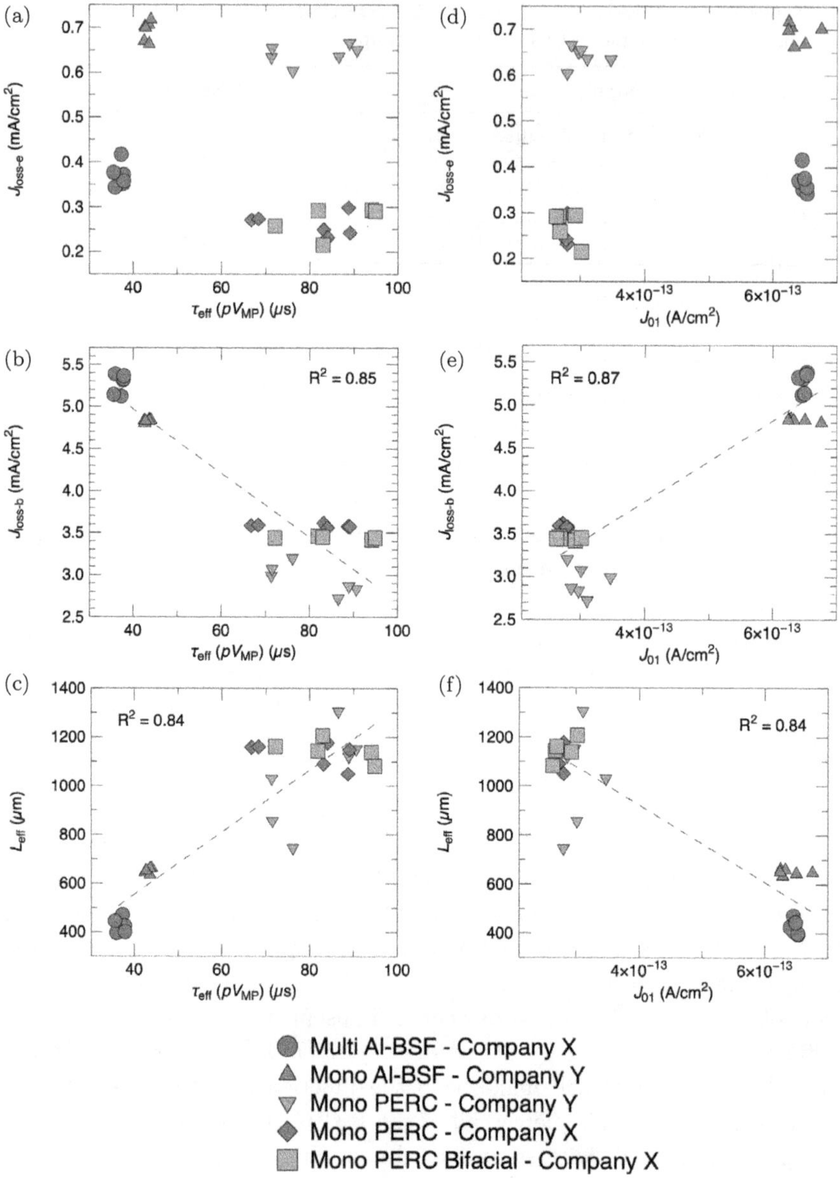

Fig. 5. (Color online) Correlation plots between (a) emitter loss ($J_{\mathrm{loss}-e}$) and the effective carrier lifetime at pseudo-maximum power voltage (τ_{eff}), (b) base loss ($J_{\mathrm{loss}-b}$) and τ_{eff} (c) Effective diffusion length in the base (L_{eff}) and τ_{eff}, (d) $J_{\mathrm{loss}-e}$ and dark saturation current density (J_{01}), (e) $J_{\mathrm{loss}-b}$ and J_{01}, and (f) L_{eff} and J_{01}.

spectral response measurements, including $J_{\text{loss}-e}$, $J_{\text{loss}-b}$, and L_{eff}. These correlations highlight the fact that the cell properties (τ_{eff} and J_{01}) are strongly correlated with the properties of the base of the device ($J_{\text{loss}-b}$ and L_{eff}). The emitter loss does not have a significant impact on the resulting cell properties. If resistance losses were investigated, we would expect to see the emitter properties having the most significant impact on the contact resistivity and sheet resistance.

3.3. *Spatially resolved losses*

Along with cell level device parameters, spatially resolved images were generated for all of the current loss parameters along with V_{OC} images generated from PL measurements. One representative cell in each group is shown in Fig. 6 for each of the spatially resolved parameters. Also included in this figure is a histogram showing the distribution of each parameter. The first observation is that the V_{OC} and J_{SC} images appear very similar to each other based on a qualitative visual inspection. In many cases, defects impacting the V_{OC} also have an impact on the J_{SC}. The other observation is that the J_{SC} images are very similar to the $J_{\text{loss}-b}$ images, indicating that $J_{\text{loss}-b}$ is the dominant loss factor with respect to current.

Most of the monocrystalline devices have very little spatial variation. One exception is the monocrystalline PERC from Company Y, where there is a large area defect near the top of the cell. This pattern, showing up in both the J_{SC} and V_{OC} images, is likely an issue with the rear side passivation layer as the $J_{\text{loss}-b}$ is the only current loss parameter affected. This artifact is also visible as a double peak in the histograms for these parameters.

For the multicrystalline Al-BSF cell, there are a few interesting observations. First, the V_{OC} and J_{SC} images appear to have a strong correlation with each other, where both images exhibit a remarkably similar grain structure. This observation emphasizes the fact that the bulk wafer quality is still one of the most significant drivers for cell performance. Another interesting observation is the nonuniform optical behavior, visible in the images for the J_{R-f} and $J_{R-\text{esc}}$. The isotexture process for multicrystalline is intended to allow similar texturing of all grains regardless of their orientation. The visibility of grains in the J_{R-f} images indicate, which grains are more easily textured than others. The histogram also shows the impact of this different texturing process as there is a substantial increase in the average along with an increase in the spatial variance. For all of the monocrystalline groups, there is a tight spatial distribution for

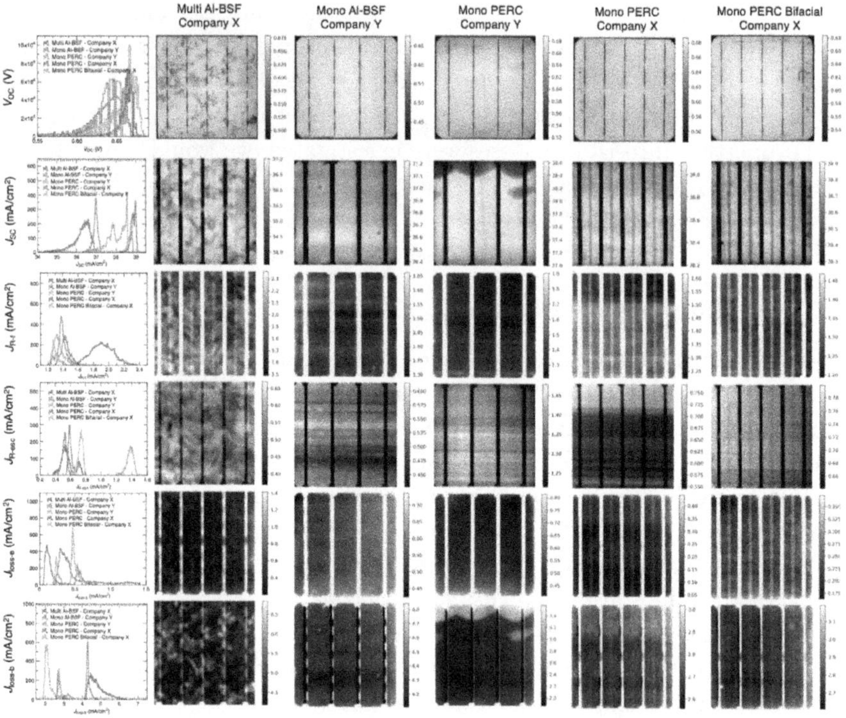

Fig. 6. (Color online) Parameter images for one representative cell from each group derived from PL images ($V_{\rm OC}$) and from spatially resolved EQE and reflectance ($J_{\rm SC}$, J_{R-f}, $J_{R-{\rm esc}}$, $J_{{\rm loss}-b}$, $J_{{\rm loss}-e}$). Histograms for each parameter are displayed to the left of all of the images.

this parameter. There are also patterns observed in the escape reflectance for the multicrystalline cells. This indicates changes in the optical properties of the rear surface that depend on grain orientation. It is interesting to note that optical properties of the front and rear surface are not entirely correlated with each other, as different patterns are visible in these two images. Finally, the relatively uniform nature of the $J_{{\rm loss}-e}$ images indicate that the diffusion process is very uniform with respect to grain orientation.

One final point of consideration is how these spatially resolved characterization techniques could be used in a manufacturing environment. Currently, PL and electroluminescence (EL) imaging are the primary methods of obtaining spatially resolved information about PV cells in research and manufacturing. Both techniques are relatively fast and can be easily performed in less than 1 S, making them applicable as in-line metrology techniques. Because the luminescence intensity of each pixel can be related to a local voltage, both methods are effective at characterizing recombination and resistive losses in 2D spanning the entire surface of the cell. Techniques and even complete software tools (e.g., Quokka, Griddler) are available to assist in the analysis of these images to gain insight into these specific losses.[74–85] EL and PL can be used for process control, to detect performance limiting problems in manufacturing, or even to detect potential reliability issues at different stages of manufacturing.[86–89]

Spatially resolved EQE and R measurements are less common, but have the ability to obtain information about both the optical and recombination losses and to distinguish the location of recombination losses in a 3D, with respect to the front, bulk, or rear of the cell. This can be very useful in terms of tracking down the source of defects (e.g., front texture, front passivation, front diffusion, front contact recombination, rear passivation, rear contact recombination) or as way to monitor upstream unit processes using data measured downstream at the cell level. The integration of spatially resolved EQE measurements into high-volume manufacturing as an in-line metrology process may require a noncontact approach (i.e., photons in, photons out). This could be accomplished in the future by measuring luminescence intensity for different wavelengths of optical excitation.[90–93]

4. Conclusions

Five unique p-type c-Si solar cell groups were investigated, including three unique cell architectures (Al-BSF, PERC, and bifacial PERC) from two manufacturers. Characterization of these cell groups were carried out using PL, *suns*-V_{OC}, and spatially resolved EQE and reflectance. An investigation into the loss mechanisms responsible for device performance was carried out. A strong correlation was observed between V_{OC} and J_{SC} indicating that the main driver influencing performance is recombination related. The only group that deviated from this trend was the multicrystalline Al-BSF group. This was linked to the use of a different texturing process that impacted the optical performance, but not recombination properties of the cell. The transition to PERC from an Al-BSF architecture was also linked to an increase in the variance from sample to sample with respect to τ_{eff}. Spatially resolved characterization was used to understand the magnitude of spatial variance across the various cell groups. As anticipated, the multicrystalline group showed the largest spatial variance, as the bulk material quality was the dominant driver of cell performance. The spatially resolved analysis also allowed for the detection of nonuniform processing. In this group of cells nonuniformities were identified in the rear passivation for one monocrystalline PERC group and in the texturing of the multicrystalline Al-BSF group.

Acknowledgments

This material is based upon work supported by the U.S. Department of Energy's Office of Energy Efficiency and Renewable Energy (EERE) under Solar Energy Technologies Office (SETO) Agreement Number DE-EE0008172.

References

1. M. A. Green, *Sol. Energy Mater. Sol. Cells* **143**, 190 (2015).
2. T. Dullweber and J. Schmidt, *IEEE J. Photovol.* **6**, 1366 (2016).
3. A. Goodrich *et al.*, *Sol. Energy Mater. Sol. Cells* **114**, 110 (2013).
4. C. Battaglia, A. Cuevas and S. De Wolf, *Energy & Environ. Sci.* (2016).
5. B. Min *et al.*, *IEEE J. Photovol.* **7**, 1541 (2017).
6. A. G. Aberle *et al.*, *Green* **2**, (2012).
7. K. Krauß *et al.*, *Physica Status Solidi A* **213**, 68 (2016).
8. R. Guerrero-Lemus *et al.*, *Renew. Sust. Energy Rev.* **60**, 1533 (2016).
9. R. S. Bonilla *et al.*, *Physica Status Solidi A* **214**, 1700293 (2017).
10. G. Dingemans and W. M. M. Kessels, *J. Vac. Sci. & Technol.* **30**, 040802 (2012).
11. B. Hoex *et al.*, *Appl. Phys. Lett.* **89**, 042112 (2006).
12. B. Hoex *et al.*, *Appl. Phys. Lett.* **91**, 112107 (2007).
13. M. Z. Rahman, *Renew. Sust. Energy Rev.* **30**, 734 (2014).
14. M. Z. Rahman and S. I. Khan, *Mater. Renew. Sust. Energy* **1**, 1 (2012).
15. P. Saint-Cast *et al.*, *IEEE Electron Device Lett.* **31**, 695 (2012).
16. P. Saint-Cast *et al.*, *Appl. Phys. Lett.* **95**, 151502 (2009).
17. L. E. Black and K. R. McIntosh, *Appl. Phys. Lett.* **100**, 202107 (2012).
18. K. O. Davis *et al.*, *Physica Stat. Solidi-Rapid Res. Lett.* **7**, 942 (2013).
19. M. Kim, S. Park and D. Kim, *Sol. Energy Mater. Sol. Cells* **117**, 126 (2013).
20. M. Kim *et al.*, *Sol. Energy* **108**, 101 (2014).
21. M. Kim *et al.*, *Sol. Energy* **110**, 208 (2014).
22. O. Breitenstein, *Opto-Electron. Rev.* **21**, 259 (2013).
23. O. Breitenstein, J. P. Rakotoniaina and M. H. A. Rifai, *Prog. Photovoltaics Res. Appl.* **11**, 515 (2003).
24. O. Breitenstein, *Sol. Energy Mater. Sol. Cells* **95**, 2933 (2011).
25. O. Breitenstein, *Sol. Energy Mater. Sol. Cells* **107**, 381 (2012).
26. O. Breitenstein *et al.*, *Solid State Phenom.* **156**, 1 (2010).

27. O. Breitenstein, F. Fertig and J. Bauer, *Sol. Energy Mater. Sol. Cells* **143**, 406 (2015).
28. O. Breitenstein *et al.*, *Sol. Energy Mater. Sol. Cells* **137**, 50 (2015).
29. F. Frühauf, J. Wong and O. Breitenstein, *Sol. Energy Mater. Sol. Cells* **189**, 133 (2019).
30. F. Frühauf, Y. Sayad and O. Breitenstein, *Sol. Energy Mater. Sol. Cells* **154**, 23 (2016).
31. F. Frühauf *et al.*, *Sol. Energy Mater. Sol. Cells* **162**, 103 (2017).
32. F. Fertig *et al.*, *Energy Procedia* **77**, 43 (2015).
33. M. Glatthaar *et al.*, *J. Appl. Phys.* **108**, 014501 (2010).
34. H. Höffler *et al.*, *J. Appl. Phys.* **115**, 034508 (2014).
35. H. Hoffler, O. Breitenstein and J. Haunschild, *IEEE J. Photovol.* **5**, 613 (2015).
36. Z. Hameiri, P. Chaturvedi and K. R. McIntosh, *Appl. Phys. Lett.* **103**, 023501 (2013).
37. Z. Hameiri and P. Chaturvedi, *Appl. Phys. Lett.* **102**, 073502 (2013).
38. K. Öğütman *et al.*, *Sol. Energy* **158**, 869 (2017).
39. M. J. Hossain *et al.*, *Sol. Energy Mater. Sol. Cells* **199**, 136 (2019).
40. M. J. Hossain *et al.*, *IEEE J. Photovol.* **9**, 1350 (2019).
41. K. Dressler *et al.*, *IEEE J. Photovol.* **5**, 70 (2015).
42. K. Öğütman *et al.*, *Sol. Energy Mater. Sol. Cells* **179**, 31 (2018).
43. S. Johnston *et al.*, *IEEE J. Photovol.* **4**, 348 (2014).
44. H. T. Nguyen *et al.*, *Solar RRL* **1**, 1700157 (2017).
45. F. Yan *et al.*, *Phys. Status Solidi-Rapid Res. Lett.* **6**, 190 (2012).
46. Y. Augarten *et al.*, *Prog. Photovoltaics Res. Appl.* (2012).
47. F. D. Heinz *et al.*, *Phys. Status Solidi-Rapid Res. Lett.* **13**, 1900076 (2019).
48. W. Kwapil *et al.*, *Sol. Energy Mater. Sol. Cells* **191**, 71 (2019).
49. B. Michl *et al.*, *IEEE J. Photovol.* **4**, 1502 (2014).
50. B. Michl *et al.*, *Prog. Photovoltaics Res. Appl.* (2012).
51. M. Müller *et al.*, *IEEE J. Photovol.* **2**, 586 (2012).
52. M. Padilla *et al.*, *Sol. Energy Mater. Sol. Cells* **120**, 282 (2014).
53. K. Ramspeck *et al.*, *Appl. Phys. Lett.* **90**, 153502 (2007).
54. C. Shen *et al.*, *Sol. Energy Mater. Sol. Cells* **123**, 41 (2014).
55. B. Hallam *et al.*, *J. Appl. Phys.* **115**, 044901 (2014).
56. M. Kasemann *et al.*, *Prog. Photovoltaics Res. Appl.* **16**, 297 (2008).
57. R. Evans *et al.*, *2014 IEEE 40th Photovoltaic Specialist Conference (PVSC)* (2014), p. 2699.
58. R. Evans and M. Boreland, *IEEE J. Photovol.* **8**, 38 (2018).
59. M. J. Kerr, A. Cuevas and R. A. Sinton, *J. Appl. Phys.* **91**, 399 (2002).
60. M. J. Kerr and A. Cuevas, *Sol. Energy* **76**, 263 (2004).
61. D. L. Young *et al.*, A new real-time quantum efficiency measurement system, in *33rd IEEE Photovoltaic Specialists Conference*, 11 May 2008 (IEEE, 2008), pp. 1–3.
62. E. J. Schneller *et al.*, *IEEE J. Photovol.* **7**, 957 (2017).
63. E. Vazsonyi *et al.*, *Sol. Energy Mater. Sol. Cells* **57**, 179 (1999).

64. S. C. Baker-Finch and K. R. McIntosh, *Prog. Photovoltaics Res. Appl.* **19**, 406 (2011).
65. S. C. Baker-Finch and K. R. McIntosh, *Prog. Photovoltaics Res. Appl.* **21**, 960 (2013).
66. S. C. Baker-Finch, K. R. McIntosh and M. L. Terry, *IEEE J. Photovol.* **2**, 457 (2012).
67. P. Campbell and M. A. Green, *J. Appl. Phys.* **62**, 243 (1987).
68. D. Kray, M. Hermle and S. W. Glunz, *Prog. Photovoltaics Res. Appl.* **16**, 1 (2008).
69. Z. C. Holman *et al.*, *J. Appl. Phys.* **113**, 013107 (2013).
70. Z. C. Holman *et al.*, *Sol. Energy Mater. Sol. Cells* **120**, 426 (2014).
71. K. O. Davis *et al.*, *IEEE J. Photovol.* **3**, 641 (2013).
72. Y. Yang *et al.*, *Sol. Energy Mater. Sol. Cells* **117**, 343 (2013).
73. J. Greulich *et al.*, *Energy Procedia* **27**, 234 (2012).
74. A. Fell, *IEEE Trans. Electr. Devices* **60**, 733 (2013).
75. D. Herrmann *et al.*, *IEEE J. Photovol.* **9**, 1759 (2019).
76. X. Yang *et al.*, *Sol. Energy Mater. Sol. Cells* **140**, 38 (2015).
77. P. Saint-Cast, D. Herrmann and H. Höffler, *IEEE J. Photovol.* **10**, 554 (2020).
78. J. Wong, in *2013 IEEE 39th Photovoltaic Specialist Conference (PVSC)* (2013), pp. 933–938.
79. V. Shanmugam *et al.*, *Solar Energy* **118**, 20 (2015).
80. J. Wong *et al.*, *IEEE J. Photovol.* **5**, 619 (2015).
81. J. Wong, R. Sridharan and V. Shanmugam, *IEEE Trans. Electr. Devices* **62**, 3750 (2015).
82. J. W. Ho *et al.*, *Energy Procedia* **150**, 21 (2018).
83. V. Shanmugam *et al.*, *Sol. Energy Mat. Sol. C.* **147**, 171 (2016).
84. M. Li *et al.*, *IEEE J. Photovol.* **8**, 6 (2018).
85. M. Li *et al.*, in *2019 IEEE 46th Photovoltaic Specialist Conference (PVSC)* (2019), p. 2769.
86. H. Seigneur *et al.*, *Renew. Sust. Energy Rev.* **59**, 84 (2016).
87. K. O. Davis *et al.*, *Renew. Sust. Energy Rev.* **59**, 225 (2016).
88. E. J. Schneller *et al.*, *Renew. Sust. Energy Rev.* **59**, 992 (2016).
89. K. O. Davis *et al.*, in *2016 IEEE 43rd Photovoltaic Specialist Conference (PVSC)* (2016), p. 3495.
90. D. Berdebes *et al.*, *IEEE J. Photovol.* **3**, 1342 (2013).
91. K. O. Davis *et al.*, in *2017 IEEE 44th Photovoltaic Specialist Conference (PVSC)* (2017), p. 3448.
92. M. K. Juhl, M. D. Abbott and T. Trupke, *IEEE J. Photovolt.* **7**, 1074 (2017).
93. A. Paduthol, M. K. Juhl and T. Trupke, *J. Appl. Phys.* **123**, 023105 (2018).

https://doi.org/10.1142/9789819818433_0025

Chapter 25

Effects of SiC Nanowires on Reliability of Sn58Bi-0.05GNSs/Cu Solder Joints[#]

Nan Jiang[*], Liang Zhang[*,†,§], Kai-Kai Xu[*], Mu-Lan Li[*] and Feng-Jiang Wang[‡]

[*]*School of Mechatronic Engineering, Jiangsu Normal University, Xuzhou 221116, P. R. China*

[†]*State Key Laboratory of Advanced Welding and Joining, Harbin Institute of Technology, Harbin 150001, P. R. China*

[‡]*School of Materials Science and Engineering, Jiangsu University of Science and Technology, Zhengjiang 212000, P. R. China*

[§]*zhangliang@jsnu.edu.cn*

In this work, SiC nanowires (SiC NWs) reinforced SBG (Sn-58Bi-0.05GNSs) composite solder was prepared using powder metallurgy route. The effect of SiC NWs on melting temperature, wetting behavior, shear properties, microstructure of the prepared solder joints and interfacial reaction were studied in detail. Results reveal that incorporating SiC NWs can develop the wetting behavior and shear properties of solder joint but has a little effect on melting temperature. The microstructure of solder is refined markedly with the addition of SiC NWs, which is one of the reasons for the increase in the shear strength of the solder joints. Additionally, the dimension of Cu_6Sn_5 IMC grains diminishes with the doping of SiC NWs, which resulted in the thinning of Cu_6Sn_5 IMC layer. Thence, the addition of SiC NWs may be an effective way to improve the reliability of solder joints.

Keywords: SiC nanowires; powder metallurgy route; grains; reliability.

1. Introduction

In the electronics industry, solder can function as the mechanical connection, signal transmission and integration system.[1,2] With the development

[§]Corresponding author.

[#]To cite this article, please refer to its earlier version published in the *International Journal of Modern Physics B*, Volume 35, 2150007 (2021), DOI: 10.1142/S0217979221500077.

of environmental awareness, some lead-free solders gradually replace the SnPb solder. In this lead-free solder, Sn-58Bi solder has the merit of inexpensive cost and relatively low melting point (139°C), which was applied to the field of low-temperature soldering.[3] However, the inherent brittleness of Bi element will result in poor ductility of SnBi solder, which will affect the reliability of solder joints in service.[4]

Presently, there are some reports on the use of nanometer materials to improve the properties of Sn-58Bi solder. Several oxide nanoparticles, such as ZrO_2, $BaTiO_3$, Al_2O_3, TiO_2, have been incorporated into Sn-58Bi solder to improve the wetting behavior of Sn-58Bi solder.[5,6] Yang *et al.*[7] studied the effect of adding Mo nanoparticles on mechanical properties of Sn-58Bi solder joint during thermal cycling. The results revealed that the tensile and shear strength were both developed via adding Mo nanoparticles. Li *et al.*[8] prepared Cu_6Sn_5 nanoparticles through coprecipitation reduction method and investigated the influence of incorporating Cu_6Sn_5 nanoparticles on microstructure of Sn-58Bi solder. It was found that the microstructure of Sn-58Bi solder was refined by adding Cu_6Sn_5 nanoparticles. The addition of Cu_6Sn_5 nanoparticles can promote the nucleation process, thereby forming the finer microstructure of solder. Xiong *et al.*[9] reported that the addition of CuZnAl nanoparticles can improve the reliability of Cu/Sn-58Bi/Cu solder joints. CuZnAl nanoparticles are apt to adsorb on the IMC surface, reduced the surface energy of IMC grains and enhanced the stability of the crystal surface.[10] Ma *et al.*[11] manufactured the Sn-58Bi composite solders by powder metallurgy route and found that the addition of graphene nanosheets (GNSs) reduced the coefficient of thermal expansion of solder during thermal aging. The decrease of the coefficient of thermal expansion retarded the expansion of solder, which reduced the presence of voids.[12] Yang *et al.*[13] reported that the addition of Carbon Nanotubes (CNTs) improved the creep resistance of Sn-58Bi/Cu solder joint. CNTs can disperse in the solder matrix, which blocked the the sliding of grain boundary and the movement of intergranular dislocation.[14] In these nanometer materials, owe to its superior mechanical, electrochemical properties, corrosion and oxidation resistance, SiC nanowires (SiC NWs) can play a crucial role in strengthening and toughening composite materials.[15] However, in previous researches, there is no report on the application of SiC NWs to strength the properties of lead-free solder.

In this paper, the SBG-xSiC NWs solder was prepared through the powder metallurgy route. To estimate the reliability of the prepared solder, the melting temperature, wetting behavior, shear properties, microstructure of solder joint and interfacial reaction were investigated comprehensively.

2. Experimental Procedures

2.1. *Materials*

Sn58Bi alloy powder with the average dimension of 25–45 μm was provided by Shanghai Yusui welding material (China). The GNSs powder with the average dimension of 1–5 μm is shown in Fig. 1(a). The SiC nanowires powder with the diameter of 100–500 nm and the length of 50–100 μm was used in this study, as revealed in Fig. 1(b).

Powder metallurgy route was used to prepare the Sn-58Bi-0.05GNSs (SBG) composite solder. The SBG solder was utilized as the matrix solder. SiC nanowires powder with varying weight fractions of 0, 0.025, 0.05, 0.075, 0.1 and 0.2 wt.% and SBG powder was weighed in electronic balance, respectively. Afterwards, mix them uniformly in V-blender. After that, the mixture was compressed and sintered in the N_2 atmosphere to get the solder ingots. Last, the ingots were incised into the small pieces, and then the solder pieces were extruded into the thin solder sheets.

2.2. *Melting test*

Differential thermal analysis (DTA) was conducted on the basis of the Japanese industrial standard JIS Z3198 to measure the melting temperature of solder. Approximately, 15 mg solder pieces were sealed in aluminum pan, and then heated to 160°C at the rate of 10°C/min.

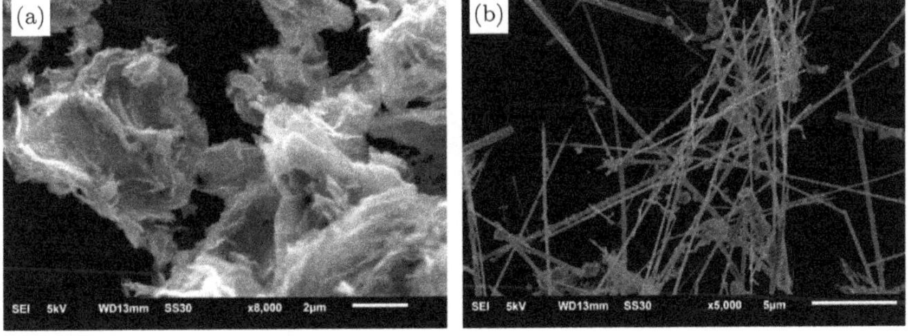

Fig. 1. SEM images of (a) GNSs and (b) SiC NWs.

2.3. *Wetting experiment*

In order to determine the solder wettability on the Cu substrate, the spreading area method was be utilized in this experiment. The size of Cu substrates was 20 mm × 20 mm × 0.3 mm. Before the test, the substrates were carried out from the ultrasonic cleaning to remove impurities. About 0.2 g solder sheets were placed on a Cu substrate with the rosin flux, and then heated to 180°C for 1 min on a graphite thermostat heating plate. The spreading area of solder on the Cu substrate was computed with Image-J software.

2.4. *Shear strength test*

Two Cu plates with the solder pieces after lapping welding formed the test sample. The dimensions of Cu plates were 40 mm × 4 mm × 0.6 mm. The solder pieces with the length of 3 mm and the thickness of 0.3 mm were used in this experiment. Figure 2 revealed the schematic diagram of test sample. Shear test was conducted by electronic universal testing machine (UTM5000) at a room-temperature. Meanwhile, the shear speed was remained unchanged at 2 mm/min.

2.5. *Microstructural and interfacial IMC layer analyzes*

The samples employed for wettability test were cross-sectioned and then embedded in epoxy. The samples in molds were taken after cooling, and then ground, polished mechanically and etched for 3 s with 5% (by volume) HNO_3 in 95% CH_3CH_2OH. The FEI Inspect S50 SEM with energy-dispersive X-ray spectroscopy (EDS) was employed to the analyzes of microstructure and interfacial IMC. The average thickness of IMC was expressed as

$$\chi = \frac{A}{L}, \tag{1}$$

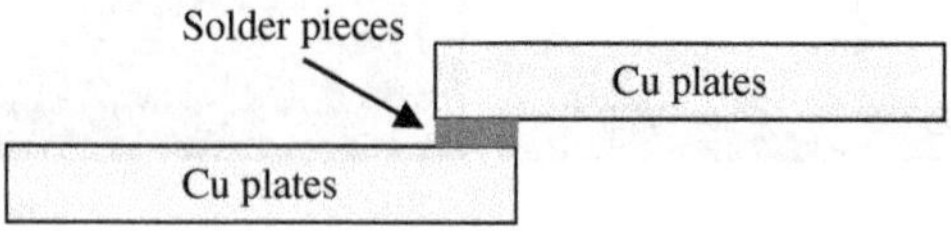

Fig. 2. (Color online) The schematic diagram of sample for shear test.

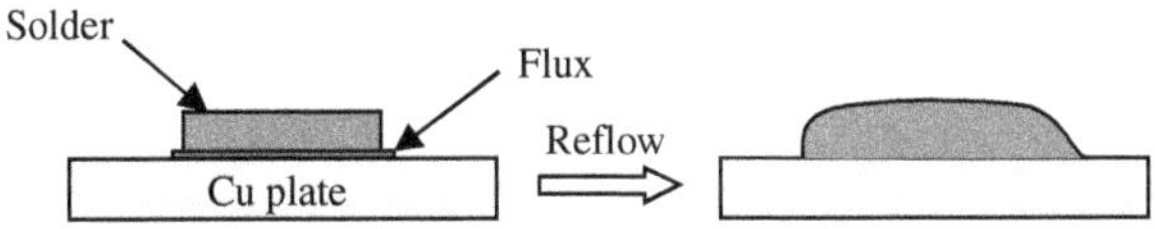

Fig. 3. (Color online) Schematic illustration of solder/Cu joints for grain analyze.

where χ refers to the thickness of IMC layer, A refers to the integral area of IMC layer, L refers to the of the IMC length.

Additionally, in order to analyze the effect of SiC NWs on IMC grains, the solder pieces were placed on a Cu plate with the rosin flux, and then heated to 200°C for 30 min on a graphite thermostat heating plate. Figure 3 shows the schematic illustration of solder/Cu joints. After reflowing, these solder/Cu joints were cleaned by Ultrasonic cleaning machine, and then ground by abrasive paper with a particle size of 400. Ultrasound oscillation was conducted in 25% (by volume) HNO_3 and 75% H_2O solution. The JSM 6510 SEM was used to observe the grain morphology of interfacial IMC.

3. Results and Discussion

3.1. *Melting characteristics*

The DSC traces of the SBG and SiC NWs added SBG solder are illustrated in Fig. 4. There was only one single endothermic peak on the DSC curve, indicating that it had an endothermic heat process. It was found that the peak temperature of solder ranged from 140.56°C to 141.35°C during the heating process. It indicated that the addition of SiC had little effect on the melting temperature of SBG solder. SiC has very high melting point (2700°C) and will not react with the matrix solder, resulting in little influence on the melting temperature of the solder. In addition, it is seen that the melting range of solder is tiny. The small melting range can reduce thermal damage during soldering process, which is beneficial to the forming process of soldering. Thence, the solder with the addition of SiC NWs has superior thermal properties, which can ensure the popularization and application of Sn-58Bi solder.

3.2. *Wettability*

Figure 5 gives the spreading area of the SBG composite solder on the Cu substrate. It is well-known that the larger spreading area means the more excellent wettability.[16] With the content of SiC NWs increased from 0 wt.%

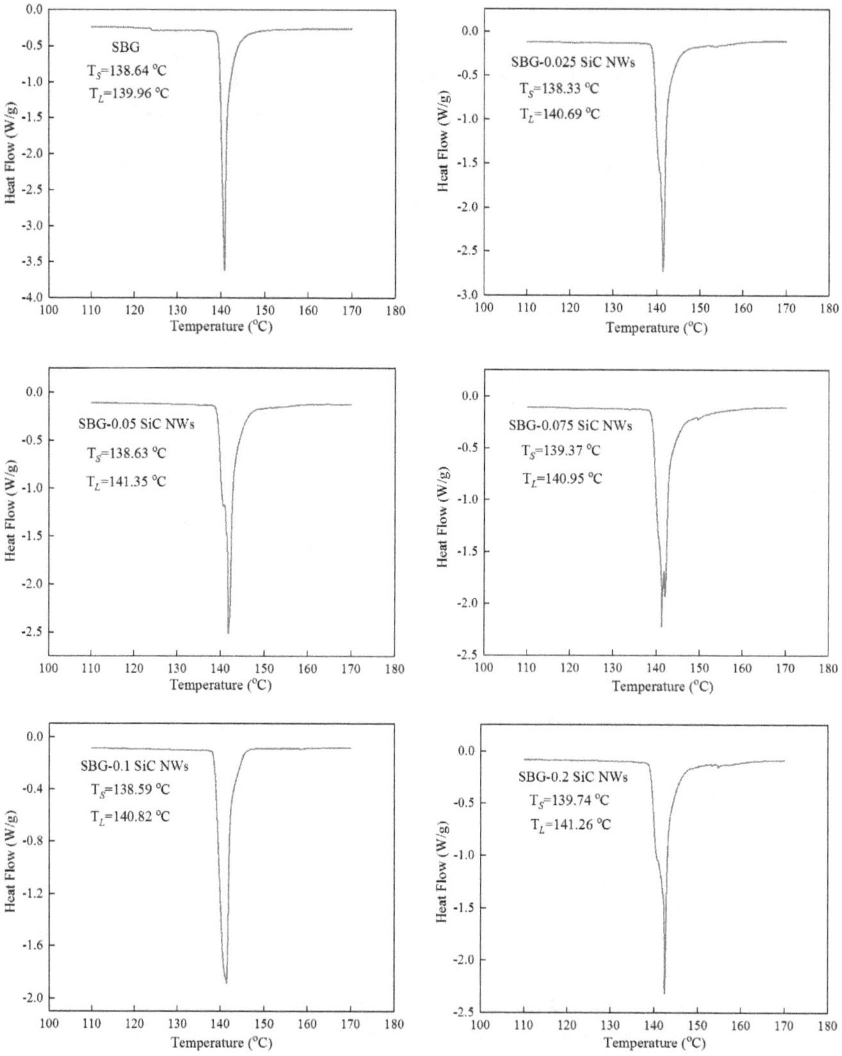

Fig. 4. (Color online) The DSC scans of solder.

to 0.05 wt.%, the solder possessed the larger spreading area. The SBG-0.05SiC NWs solder showed the largest spreading area, which was 18.9% higher than that of the pure solder. The results revealed that the addition of SiC NWs markedly improved the wettability of solder on the Cu substrate. It is well-known that the capillary phenomenon also exists at the solder/Cu

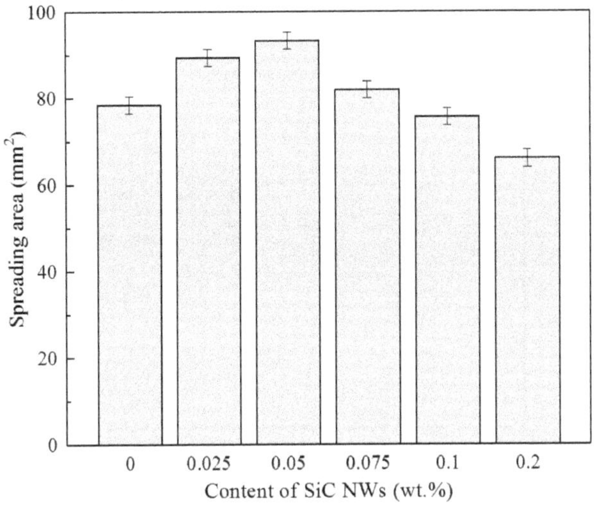

Fig. 5. (Color online) Spreading area of SBG solder and its composite solders on Cu substrate.

interface and is benefit to the improvement of solder fluidity.[17] SiC NWs are not melt during the soldering process, and can act as the reinforcing phase, which strength the capillary phenomenon. This explained why adding SiC NWs improved the wetting behavior of solder.

However, when the content of SiC NWs exceeded 0.05 wt.%, the spreading area of solder deteriorated significantly. Especially, the SBG-0.2SiC NWs solder alloy exhibited the smallest spreading area, indicating that introduction of excess SiC NWs can deteriorate the wetting performance of solder. This decrease can be related to the reunion phenomenon of SiC NWs. With the increase of additive content, the SiC NWs produced clusters under the action of van der Waals force, which eventually led to the reduction of fluidity solder. In light of the analysis, it is concluded that the optimal amount of SiC NWs was about 0.5 wt.%.

3.3. *Shear properties*

For lead-free solders, the shear properties are the vitual factor in solder joints testing.[18] Figure 6 displays the shear strength of SBG and SiC NWs-containing composite solders. It is seen that the addition of SiC NWs had the significant enhancement in the shear strength of solder joints.

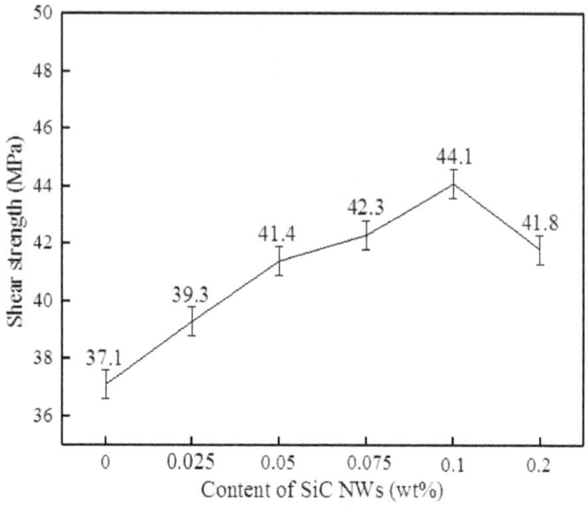

Fig. 6. Shear strength of the SBG-xSiC NWs solder joints.

The improvement of shear strength in these solder joints was 5.9–18.9% over that of the matrix solder.

The reason why the shear strength of solder joints is obviously strengthed by adding SiC NWs can be simply explained in what follows. On the one hand, SiC NWs can act as second phases particles and distribute dispersively in the solder matrix, which deflected the dislocation movement and grain-boundary sliding. According to Orowan mechanism, the dispersion strengthening can be expressed as follows:

$$\sigma_{\text{org}} = \frac{2Gb}{\lambda}, \tag{2}$$

where σ_{org} refers to the yield stress, G is the shear modulus of solder matrix, b represents the Burger's vector of the dislocation, λ refers to the distance between the dispersed SiC NWs. The yield stress improved for SiC NWs added solder alloy because of λ decrease.

On the other hand, the addition of SiC NWs impeded the migration of grain boundaries inhibited the growth of grain, which made the solder obtain the refiner microstructure. Based on Hall-Petch mechanism, the refinement strengthening equation can be shown as follows:

$$\sigma = \sigma_0 + Kd^{-1/2}, \tag{3}$$

where σ refers to the yield strength, σ_0 refers to the yield strength of single crystal; K refers to the material constants; d refers to the grain size. According to Eq. (3), it is concluded that the strength developed with the reduce of grain size in solder matrix.

3.4. *Microstructure*

SEM images in Fig. 7 show the microstructure of SBG solder and composite solders. It is well-known that the microstructure of the Sn-58Bi solder expressed a representative eutectic structure with Sn-rich phase and Bi-rich phase. The gray Sn-rich phase interlocked alternately with white Bi-rich phase, which resulted in the lamellar structure of Sn-58Bi solder.[19] A similar situation occurs in the SBGs solder, as shown in Fig. 7(a). With the dope of SiC NWS, some strip Bi phase and spheroid Bi phase appeared in the microstructure of solder. In addition, it is clear seen that the average inter-phase spacing diminished with the addition of SiC NWs. This demonstrated that adding SiC NWs can obtain the refiner the microstructure. Especially, when the content of SiC NWS was 0.1 wt.%, the inter-phase spacing was relatively small. SiC NWs can pin at grain boundaries, which impeded the migration of grain boundaries inhibited the growth of grain.[20] However,

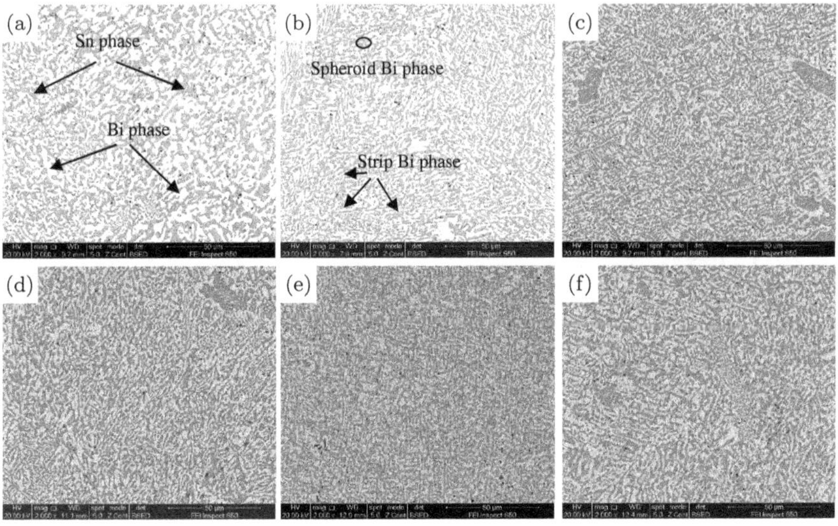

Fig. 7. Microstructure of SBG-xSiC NWs solders: (a) $x =$, (b) $x = 0.025$, (c) $x = 0.05$, (d) $x = 0.075$, (e) $x = 0.1$ and (f) $x = 0.2$.

when the content of SiC NWS was 0.2 wt.%, the microstructure of solder coarsened markedly. The coarsening of the microstructure may be caused by the agglomeration of SiC NWs under the action of van der Waals force.

3.5. *Interfacial IMC layer*

Figure 8 reveals the IMC layer morphologies of prepared solders with different content of SiC NWs. A layer of scallop-shaped IMC was formed at the SBG solder/Cu interface, as shown in Fig. 8(a). The component of interfacial IMC was determined as Cu_6Sn_5 with the EDS analysis. The formation of Cu_6Sn_5 can be expressed as

$$6Cu + 5Sn \rightarrow Cu_6Sn_5. \tag{4}$$

During the reaction, 6 mol Cu reacted with 5 mol Sn formed 1 mol Cu_6Sn_5. With the SiC NWs added, the interfacial Cu_6Sn_5 IMC became relatively flat.

The IMC thickness was measured by Image-J software so as to further analyze the effect of SiC NWs on the interfacial IMC. The IMC thickness

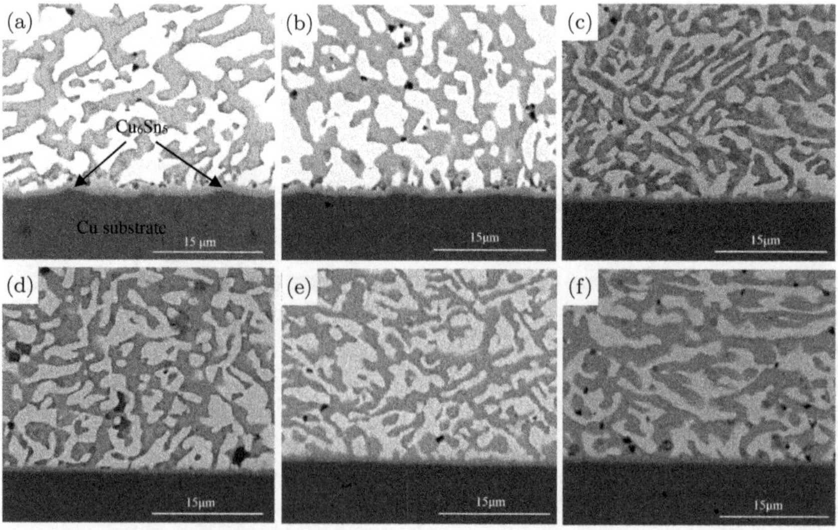

Fig. 8. SEM images of interfacial IMC layer: (a) SBG, (b) SBG-0.025SiC NWs, (c) SBG-0.05SiC NWs, (d) SBG-0.075SiC NWs, (e) SBG-0.1SiC NWs and (f) SBG-0.2SiC NWs.

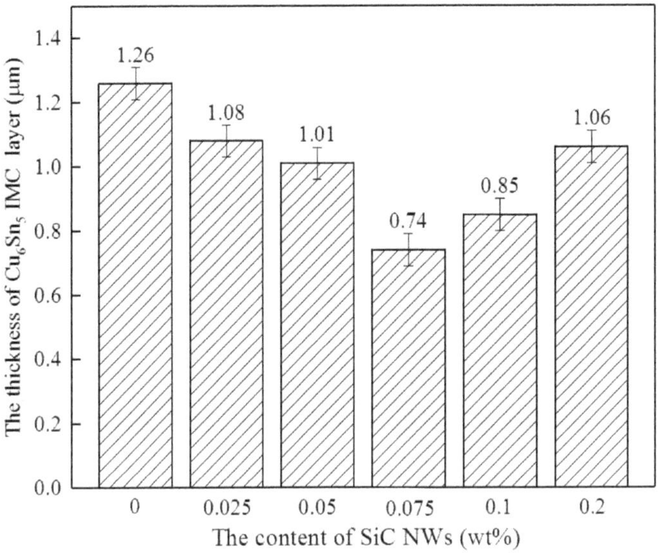

Fig. 9. The IMC layer thickness of SBG solder.

of solder/Cu joints is exhibited in Fig. 9. It is seen that the Cu_6Sn_5 IMC can be reduced with the increase of SiC content. When the content of SiC NWs increased to 0.075 wt.%, the thickness of Cu_6Sn_5 IMC is greatly decreased to 0.74 μm. However, the thickness of Cu_6Sn_5 IMC developed with the further addition of SiC NWs. When the the content of SiC NWs was 0.2 wt.%, the Cu_6Sn_5 thickness of solder was still lower than that of the SBG solder. This revealed that the addition of SiC NWs could inhibit the growth of interfacial IMC. From the point of view of solder joints reliability in service, the strength of solder joints deteriorate significantly with increasing thickness of the interfacial IMC layers[21] because of the IMC's brittleness, the crack can prorogate along the IMCs layers.[22] Adding SiC may improve the reliability of solder joints during service.

In order to explain the inhibition mechanism of SiC NWs, the Cu_6Sn_5 grains formed at solder/Cu interface was studied statistically. Figure 10 shows the SEM images of the top-view IMC in these solder/Cu joints with various content of SiC NWs. It is found that the dimension of Cu_6Sn_5 grains decreased with the introduce of SiC NWs. Especially, the Cu_6Sn_5 grains in SBG-0.1 SiC NWs solder were smaller and more uniform, which reveal that the addition of SiC NWs can block the grain ripening.

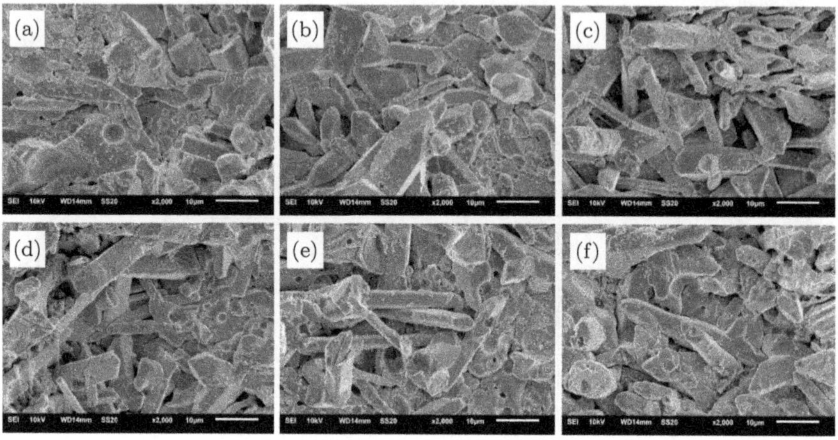

Fig. 10. SEM images of the top-view IMC in these solder/Cu joints with various content of SiC NWs.

The inhibition of grain can be explained by the adsorption effect of SiC NWs. The SiC NWs have the higher surface energy and are apt to form clusters. In light of adsorption theory, the surface free energy of Cu_6Sn_5 grains can be expressed as

$$\sum_K \gamma_c^K A_K = \sum_K \left(\gamma_0^K - RT\int_0^c \frac{\Gamma^K}{c} dc\right) A_K, \tag{5}$$

$$= \sum_K \gamma_0^K A_K - RT\int_0^c \frac{\Gamma^K}{c} dc A_K \to \min. \tag{6}$$

Here, c refers to the concentration of SiC NWs; γ_c^K refers to Cu_6Sn_5 grains with adsorption of SiC NWs; A_K refers to the area of Cu_6Sn_5 grains; γ_0^K refers to the surface tension of Cu_6Sn_5 grains without adsorption of SiC NWs; R refers to the gas constant; T refers to the absolute temperature; r^K refers to the adsorption capacity of SiC NWs at Cu_6Sn_5 grains.

Provided that the bulk of Cu_6Sn_5 grains is constant, the surface energy of the Cu_6Sn_5 planes required to keep the crystal plane minimum in equilibrium state. Thence,

$$\sum_K A_K \int_0^c \frac{\Gamma^K}{c} dc \to \max. \tag{7}$$

This means that when the adsorption amount of SiC NWs reaches the maximum value, the surface energy of the entire crystal is the lowest. Based on the Gibbs adsorption theory, it can be seen that the decrease in the surface energy of the grain will cause the decrease in the growth rate of the grain.[23]

4. Conclusions

The effect of SiC NWs on melting temperature, wetting behavior, shear properties, microstructure of the prepared solder joint and interfacial reaction was studied in detail. The main conclusions can be summarized as follows:

(1) The introduction of SiC developed markedly the wetting behavior and shear property of solder. However, the addition of SiC nanowires had little effect on the melting temperature of solder.
(2) The SiC-doped solder can obtain the refiner microstructure, and the best content of SiC was 0.075 wt.%.
(3) Compared to the SBG solder joint, the IMC thickness of solder joints with the addition of SiC NWs was significantly thinner.

Acknowledgment

This work was supported by the Key project of State Key Laboratory of Advanced Welding and Joining (AWJ-19Z04).

References

1. C. H. Du *et al.*, *Adv. Mater. Res.* **337**, 526 (2011).
2. M. Y. Xiong and L. Zhang, *J. Mater. Sci.* **54**, 1741 (2019).
3. N. Jiang *et al.*, *J. Mater. Sci.: Mater. Electron.* **30**, 17583 (2019).
4. N. Jiang *et al.*, *Sci. Technol. Adv. Mater.* **20**, 876 (2019).
5. L. Yang *et al.*, *J. Electron. Mater.* **44**, 2473 (2015).
6. T. W. Hu *et al.*, *Microelectron. Reliab.* **55**(8), 1226 (2015).
7. L. Yang *et al.*, *Mater. Charact.* **148**, 280 (2019).
8. X. Li *et al.*, *Mater. Sci. Eng.: A* **684**, 328 (2017).
9. M. Y. Xiong *et al.*, *Vacuum* **167**, 301 (2019).
10. F. Yang *et al.*, *Materials* **10**, 558 (2017).
11. Y. Ma *et al.*, *Mater. Sci. Eng.* **696**, 437 (2017).
12. H. G. Wang, K. K. Zhang and M. Zhang, *J. Alloys Compd.* **781**, 761 (2019).
13. L. Yang, H. X. Liu and Y. C. Zhang, *J. Electron. Mater.* **47**, 662 (2017).
14. H. Y. Sun, Y. C. Chan and F. S. Wu, *Mater. Sci. Eng.: A* **656**, 249 (2016).
15. C. Y. Cheng *et al.*, *J. Am. Ceram. Soc.* **101**, 3694 (2018).

16. M. Zhao *et al.*, *Sci. Technol. Adv. Mater.* **20**, 421 (2019).
17. X. J. Yao, Z. D. Wang and W. J. Zhang, *IEEE Trans. Compon. Packag. Manuf. Technol.* **4**, 1534 (2017).
18. L. Zhang and Z. Q. Liu, *J. Mater. Sci.: Mater. Electron.* **31**, 2466 (2020).
19. L. Zhang, L. Sun and Y. H. Guo, *J. Mater. Sci.: Mater. Electron.* **26**, 7629 (2015).
20. X. H. Shi *et al.*, *J. Alloys Compd.* **693**, 837 (2016).
21. F. J. Wang *et al.*, *J. Mater. Sci.: Mater. Electron.* **29**, 11409 (2018).
22. M. Zhao *et al.*, *Int. J. Mod. Phys. B* **34**, 2050064 (2020).
23. L. Sun, M. H. Chen and L. Zhang, *J. Mater. Process. Technol.* **278**, 116507 (2020).

https://doi.org/10.1142/9789819818433_0026

Chapter 26

Insightful Study of the Characterization of the Cobalt Oxide Nanomaterials and Hydrothermal Synthesis[#]

Muhammad Yasir Raza*, Jamil Abbas Haider†, N. Ameer Ahammad‡, Kamel Guedri§ and Ahmed M. Galal¶,‖,**

*Department of Physics, Bahauddin Zakariya University, Multan 60800, Pakistan

†Abdus Salam School of Mathematical Sciences, Government College University, Lahore 54600, Pakistan

‡Department of Mathematics, Faculty of Science, University of Tabuk, P. O. Box 741, Tabuk 71491, Saudi Arabia

§Mechanical Engineering Department, College of Engineering and Islamic Architecture, Umm Al-Qura University, P. O. Box 5555, Makkah 21955, Saudi Arabia

¶Department of Mechanical Engineering, College of Engineering in Wadi Alddawasir, Prince Sattam bin Abdulaziz University, Saudi Arabia

‖Production Engineering and Mechanical Design Department, Faculty of Engineering, Mansoura University, P. O. 35516, Mansoura, Egypt

**ahm.mohamed@psau.edu.sa

In this paper, nanoparticles of cobalt oxide (Co_3O_4) are prepared at different temperatures 120°C, 140°C and 160°C using the hydrothermal method. Cobalt nitrate hexahydrate: Co $(NO_3)_2 \cdot 6H_2O$ is used as precursor and potassium hydroxide (KOH) is used as precipitating agent. Particle size is controlled using precursor concentration. It is also investigated in this research that particle size increases at high-temperature. Nanoparticles of size between (13.62–17.81 nm) are obtained using this technique (Hydrothermal method). SEM results provide nonuniform distribution of nanoparticles with sharp grain boundaries. Electrical characterization confirms the semiconducting

**Corresponding author.

#To cite this article, please refer to its earlier version published in the *International Journal of Modern Physics B*, Volume 37, 2350101 (2023), DOI: 10.1142/S0217979223501011.

behavior of the material as resistivity decreases with increase in temperature. Electrochemical measurements show detection of hydrogen peroxide H_2O_2 by nanoparticles of Co_3O_4.

Keywords: Nanomaterials; cobalt oxide (Co_3O_4); temperature; hydrothermal method.

1. Introduction

Oxides and compounds of oxides may be found in a number of semiconductors. The amount of oxygen present in these substances is linked to their individual physical characteristics. There is a wide range of variation in their electrical properties due to the different oxygen contents. The oxygen concentration of these materials is rather high, which contributes to their high conductivity.

Materials with a high oxygen concentration have a lot of oxygen in them, which is why they don't conduct electricity well.[1] The stable oxides of cobalt such as CoO and Co_3O_4 are used in sensing of toxic gases in the environment and other useful application in the industry.[2–4] Bulk cobalt oxide has improved the crystal structure and grain size which induce higher conductivity. The grain size of cobalt oxide is influence by annealing. Annealing effect is observed from the results of direct bandgap energy and electrical resistivity of Co_3O_4.[5] Due to higher conductivity, Co_3O_4 electrodes have been used to investigate the electrochromic devices.[6] Co_3O_4 material is cathode catalyst for fuel cells using H_2O_2 as oxidant. Co_3O_4 is stable for electro-catalytic reduction of H_2O_2.[7] Co_3O_4 can be used as sensing device for H_2O_2. Hydrogen peroxide reduction produces high current density.[8] Co_3O_4 is *p*-type semiconductor with properties of sensing,[9] fuel cell electrode[10] and electrochemical properties, and has been used for applications in solid-state sensors, electrochromic devices and lithium batteries.[11] In gas sensor, gas molecules interact chemically with molecules of metal oxide on the sensor surface. Efficient sensors are influence by surface area, nanoparticles size of metal oxide and thickness of electrode of sensors. Hence, the structural parameters of sensor materials, such as pore size, specific surface area, and film thickness, can greatly influence the properties of sensors.[12,13] The electrochemical properties of Co_3O_4 display super-capacitive behavior.[14] Nanoparticles of cobalt oxide have application in batteries, heterogeneous catalysts, supercapacitors and electrochemical sensors due to its grain size and shape.[15–19]

Different nanoscale morphologies, such as hollow nanospheres, cubic single crystals, fibers,[20] particles, rods,[21] tubes[22] and films in nanoscale, have been reported.[23] Different morphologies of nanoparticles have different potential for sensing chemicals. For example, hollow nanospheres of Co_3O_4 show good sensing performances towards toluene and acetone vapors high sensitivity at low operating temperature.[24] Optical properties of Co_3O_4 nanoparticles have two energy bandgaps, which confirm semiconducting properties of the oxide.[25] Research analysis demonstrates that Co_3O_4 nanowalls have high potential to use in biosensor applications.[26,27]

The purpose of this work is to investigate the morphological and structural characteristics of samples of cobalt oxide that have been generated hydrothermally. The samples in question are Co_3O_4. The findings of the hydrogen peroxide H_2O_2 detection are obtained after the nanoparticles of cobalt oxide Co_3O_4 have been placed on the electrode. Cyclic voltammetry was used in order to investigate the electrochemical characteristics.

2. Methodology

Hydrothermal processing is used to produce cobalt oxide Co_3O_4 nanoparticles. This process is carried out at temperatures of 120°C, 140°C and 160°C, respectively, and utilizes a variety of preparation methods, depending on the amount of precursor and precipitating agent used. According to the information in Table 1, the samples were given the names A, B, and C.

Methods used in the creation of each sample are detailed in Table 1. Different proportions of the precursors to the precipitant were used to construct the samples. These samples were made at different temperatures. Two of the samples analyzed by XRD were A (30 mL of precursor and 30 mL of precipitating agent) and C (which consists of 50 mL of precursor and 10 mL of precipitating agent). These two were synthesized at temperatures of 140°C and 160°C, respectively.

Table 1. Variable stoichiometric sample size.

Sample of code	A	B	C
Precursor–Volume (mL)	30.0	40.0	50.0
P. Agent-Volume (mL)	30.0	20.0	10.0
Total-volume (mL)	60.0	60.0	60.0

3. XRD Analysis

Table 1 outlines the synthesis methods for each sample. The samples were created using a variety of mixes of the precursor to the precipitating agent. At a range of temperatures, these samples were prepared. A (with a 30 mL preservative and a 30 mL precipitating agent) and C are two of the samples that were analyzed using the XRD method (It has a volume of 50 milliliters of the precursor and 10 milliliters of the precipitating agent). Both of these samples were synthesized at 140°C and 160°C.[29,30]

In Fig. 1, we compare the XRD pattern of a hydrothermally generated Co_3O_4 sample to the typical crystallographic structure of spinel cobalt oxide derived from JCPDS card 00-043-1003. There are just a few peaks visible because of the Co_3O_4 reflections.[28] We were able to calculate the typical sample size of 14.21 nm using Scherrer's formula.

A hydrothermal process at a temperature of 160°C was used to create the XRD pattern for sample A. JCPDS card 00043-1003 was used to compare the XRD pattern findings with the previous one. There existed a perfectly pure cubic phase, as shown by these observations.[28] On average, the crystallites were 14 nm in diameter. Here, the lattice constant is 8.06.[30]

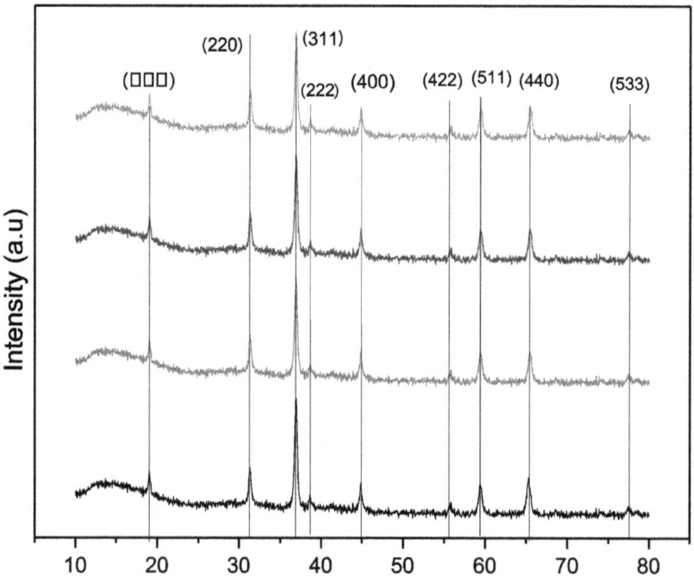

Fig. 1. (Color online) All samples' XRD graphs plotted together.

Table 2. Particle size of sample Co_3O_4 at scheme (A).

2θ	θ	FWHM(β)	hkl	particle size (nm)	Lattice constant
31.330	15.660	0.01040	(220)	13.740	8.060
36.940	18.470	0.01390	(311)	10.460	8.060
38.630	19.310	0.01320	(222)	11.070	8.060
44.840	22.420	0.00970	(400)	15.340	8.070
55.870	27.930	0.00970	(422)	16.050	8.050
59.440	29.720	0.01390	(511)	11.420	8.070
65.290	32.640	0.01460	(440)	11.220	8.070
77.480	38.740	0.01180	(533)	14.970	8.060

Table 3. Particle size of sample Co_3O_4 at 140°C (Scheme C).

19.020	9.510	0.00970	(111)	14.370	8.070
31.330	15.660	0.01040	(220)	13.740	8.060
36.90	18.450	0.01740	(311)	8.370	8.060
38.630	19.310	0.01110	(222)	13.140	8.060
44.840	22.420	0.01110	(400)	13.420	8.070
55.870	27.930	0.00760	(422)	20.420	8.050
59.440	29.720	0.01390	(511)	11.420	8.070
65.450	32.720	0.01390	(440)	11.790	8.050
77.480	38.740	0.00830	(533)	21.210	8.060

Table 4. Particle size of sample Co_3O_4 at 160°C (Scheme A).

19.02	9.51	0.0062	(111)	22.36	8.07
31.33	15.66	0.0104	(220)	13.74	8.06
36.90	18.45	0.0167	(311)	8.72	8.06
38.63	19.31	0.0153	(222)	9.56	8.06
44.84	22.42	0.0104	(400)	14.31	8.07
55.87	27.93	0.0076	(422)	20.42	8.05
59.44	29.72	0.0153	(511)	10.39	8.07
65.45	32.72	0.0174	(440)	9.43	8.05
77.48	38.74	0.0104	(533)	16.96	8.06

Samples B and C were synthesized at 160°C. The XRD pattern values were compared using JCPDS card 00-043-1003. Tables 3 and 4 using these findings, we may conclude that the pure cubic phase was present. There were an average of 17.81 nm crystals in each sample. Here too, 8.06 is the lattice constant.[27,29]

Figure 1 displays the X-ray diffraction patterns of materials A and C at temperatures of 140°C and 160°C, respectively. Nine peaks may be seen in the pattern made by the calcined product. These peaks have two values that are as follows: 19.02, 31.33, 36.94, 38.63, 44.84, 55.87, 59.44, 65.29 and 77.48. These values correspond to the crystal planes (111), (220), (311), (222), (400), (422) and (511). The diffraction patterns of the cubic structure of Co_3O_4 may be indexed to all of these different patterns. When compared to the standard values, there was almost little movement in the peaks that could be seen. This unearthed a cubic cobalt oxide structure denoted by the chemical formula Co_3O_4. Table 5 it has been determined that the value of the lattice constant, on average, is 8.06.[30–33] It is clear from looking at the image that the design of the manufactured sample is in good agreement with the generally known crystallographic configuration of cobalt oxide, which is written as Co_3O_4. Due to the fact that there are no additional peaks present, it may be deduced that the pure cubic phase formed as a result of the experimental circumstances. It is important to note that the CoO phase is absent from this system. These results demonstrated that the sample that was created has just a single phase of nanomaterial and does not include any contaminants. Sherer's equation is used in order to determine the average crystallite size of the samples.

When doing a hydrothermal synthesis, the particle size will typically grow as the reaction temperature rises. Schemes A and C were used to make nanoparticles at 140°C and 160°C, respectively. This clearly shows Table 6 as the temperature of the reaction goes up, the particle size goes up because the nucleus size expands too much at higher temperatures.[30–33]

Another tendency, this one involving an increase in particle size is also seen in this body of studies. At temperatures of 160°C and 140°C, respectively, Schemes C and A produce nanoparticles with sizes of 17.810, 14.201, 140 nm and 13.602 nm. This is due to the fact that strategy C

Table 5. Particle size of sample Co_3O_4 at 160°C (Scheme C).

19.020	9.510	0.00620	(111)	22.360	8.070
31.330	15.660	0.01040	(220)	13.740	8.060
36.90	18.450	0.01600	(311)	9.090	8.060
38.630	19.310	0.00550	(222)	26.290	8.060
44.840	22.420	0.01110	(400)	13.420	8.070
55.870	27.930	0.00690	(422)	22.470	8.050
59.440	29.720	0.01460	(511)	10.880	8.070
65.450	32.720	0.01600	(440)	10.260	8.050
77.480	38.740	0.00550	(533)	31.810	8.060

Table 6. Particle size of Co_3O_4 at different temperature and different scheme.

Sample for scheme	Temperature in °C	Particle size in (nm)	Lattice constant (Å)
(Sample-A)	(140.0°C)	(13.620)	(8.060)
(Sample-A)	(160.0°C)	(14.0)	(8.060)
(Sample-C)	(140.0°C)	(14.210)	(8.060)
(Sample-C)	(160.0°C)	(17.810)	(8.060)

makes use of a much higher quantity of the precursor than does Scheme A. It is essential to keep in mind that the mechanism known as "nucleation and growth"[34] describes the process by which nanoparticles develop when they are suspended in liquid. The nucleation process takes place over a very short period of time. The expansion of nanoparticles is directly related to the expansion of their nuclei, which may occur as a result of precipitation, diffusion, or agglomeration. When compared with a low concentration of the precursor, a high concentration results in a high diffusion rate. It is the cause of the larger particle size and the quicker development rate.[35] Using a lesser overall quantity of precipitating agent results in bigger nanoparticles compared to using a higher concentration of precipitating agent. The nucleation process consumes a significant proportion of the precursor when the precipitating agent concentration is raised. Additionally, a decrease in the quantity of precursor available for crystal formation leads in smaller nanoparticles.[36]

4. Electronic Microscopy Using SEM

For the Co_3O_4 nanoparticles made utilizing schemes A and C at 140°C and 160°C, grain size distributions are presented in Fig. 2. Specimens are examined using scanning electron microscopy (SEM) to study their morphology. At high temperatures, the grain boundary of nanoparticles is very crisp, and the particle size is extremely well defined. Based on the SEM image, the samples seem to have a semi-spherical shape. The grain boundaries of Co_3O_4 nanoparticles created at a low-temperature are clearly visible in the image. The dispersion of nanoparticles is not uniform across the sample.

4.1. *Electrical qualities that change with temperature*

The electrical properties of the material were studied using a synthetic sample in the shape of a hard pallet. Ampere values of current were determined by measuring them between zero and 10 voltages using the

Fig. 2. SEM images of Co_3O_4 at 140°C and 160°C.

source meter 2400 and the Lab tracer. To investigate the relationship between voltage and current, graphs with different temperatures were made (300 K, 333 K, 363 K, 393 K, 423 K, 453 K). Using the slope values computed from these curves, the sample's resistance (R) and resistivity (ρ) (ohm-cm) were determined.

When a sample is manufactured using hydrothermal methods, its IV curve shows that oxide Co_3O_4 nanoparticles' semiconducting properties are confirmed by this behavior Fig. 3.[29]

Activation Energy of the sample was determined using Arrhenius equation, $\rho = \rho_o \exp(Ea/\mathrm{KT})$.

Boltzmann's constant K is equal to 8.617E-5 eVK^{-1}, where ρ is the resistivity at room-temperature, Ea is the activation energy, T is the temperature. The activation energy may be calculated by finding the slope of the graph between log (ρ) (Y-axis) and 1000/T (X-axis). In order to determine the activation energy of the sample, it was subjected to both low and high voltages. When applied with a large voltage, the activation energy drops from 0.394 eV to 0.281 eV Fig. 4.

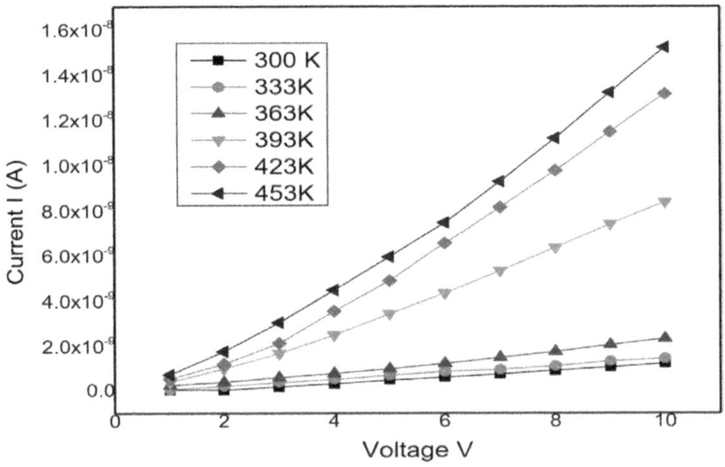

Fig. 3. (Color online) Temperature-dependent IV-curve of Sample A.

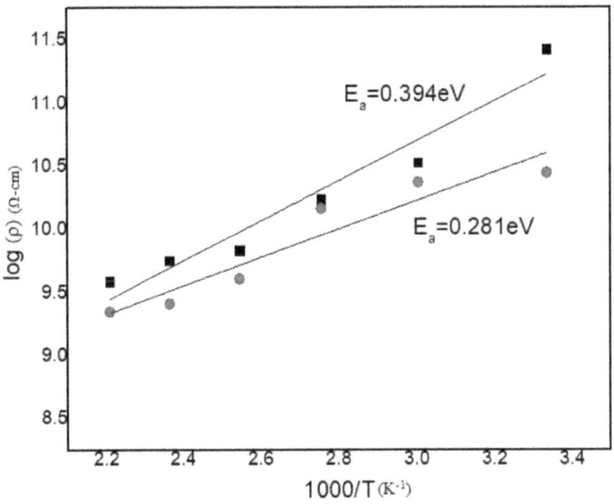

Fig. 4. (Color online) Log (ρ) resistivity (Ω-cm) versus 1000/T (K^{-1}).

4.2. *Electrochemical measurements*

In order to operate the equipment called Zennium/ZAHNER-Elektrik, which was utilized in the investigation of electrochemical measurement, the researchers made use of the software Thales Z 2.0. In order to conduct electrochemical measurements, a setup that comprises of three electrodes was used. The following is a list of the three electrodes along with their separate portfolios:

(1) Ag/AgCl (3 M NaCl) as reference electrode.
(2) A working electrode composed of modified ITO glass with the size of 2 centimeters by 1 centimeter.
(3) The counter electrode is made of platinum wires.

In order to prevent catecholamine from being oxidized, the freshly manufactured electrolyte (0.1 M PBS) was de-aerated by bubbling a steady stream of cleansed nitrogen gas. Using amperometric (A) and cyclic voltammetry (CV) methods, it was possible to examine the electro catalytic sensing capabilities of the electrode that was constructed.

4.2.1. *Electrocatalytic performance of the designed electrode*

CV measurements taken at 100 mVs^{-1} allowed us to evaluate the electrocatalytic performance of electrodes modified with cobalt oxide Co_3O_4 nanoparticles in a buffer solution with a pH of 7.4. (Fig. 3). It has been observed that the electrolyte exhibits a set of different redox peaks. When compared to this, the hydrogen peroxide H_2O_2 catalytic oxidation efficiency of cobalt oxide Co_3O_4 nanoparticles adapted ITO electrodes was further enhanced in terms of high sensitivity and current density with the addition of 1 mM of H_2O_2 in a buffer solution with a pH of 7.4 at 100 mVs^{-1}.

4.2.2. *Measurements using electrochemical sensing*

In order to Figs. 5 and 6 the sensing capabilities of the electrodes that were manufactured, amperometric measurements were carried out to identify the different concentrations of hydrogen peroxide H_2O_2 that were used. This was done so that the sensing capabilities of the electrodes could be demonstrated. The image depicts the settings that determine the amperometric current response of Co_3O_4 nanoparticles reconstructed electrodes to 50 nM H_2O_2 injections at 0.33 V (versus Ag/AgCl) in 0.1 M N2-saturated PBS (pH 7.4). The amperometric current response gradually grew when H_2O_2 was continually provided, and after 3 s, it had reached 96% of the steady state current. This was the case when the steady state current was maintained. As shown in the plot between current density and concentration (Fig. 6), the detection limit of the cobalt oxide Co_3O_4 nanoparticles centered electrode toward hydrogen per oxide H_2O_2 was revealed to be 0.1 nM. This information is shown in the graph.

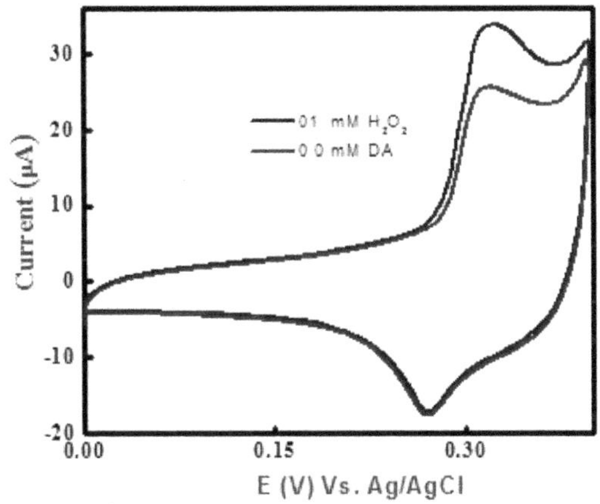

Fig. 5. (Color online) Typical CV graphs of Co_3O_4 nanoparticles based electrode in a buffer solution (pH = 7.4) without (red line) and with (blue line) 1 mM of H_2O_2 at a scan rate of 100 mV/s.

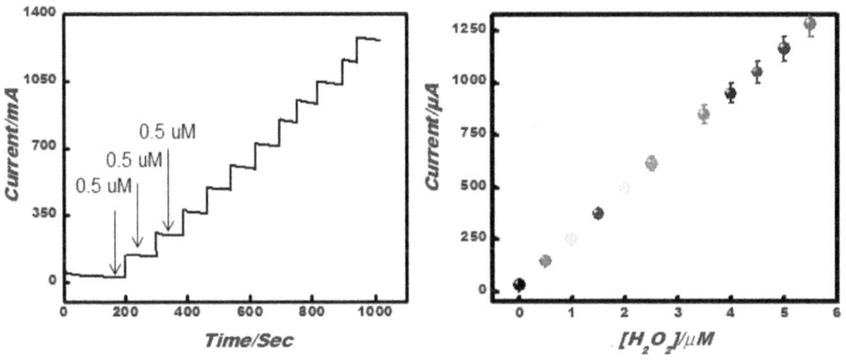

Fig. 6. (Color online) Co_3O_4 nanoparticles-based electrode amperometric curve for sequential addition of H_2O_2 in 0.1 M PBS (pH = 7.4) at 0.33 V. (b) Current-versus-time graphic yields calibration curve.

5. Concluding Remarks

The hydrothermal approach is used to create Co_3O_4 nanoparticles for the purpose of this investigation as follows:

(1) The sample that was obtained using this procedure is entirely of the cubic phase. Nanoparticles of Co_3O_4 have a crystal size determined by XRD to be between 13.62 and 17.81 nm.

(2) A rise in the concentration of the precursor as well as the temperature causes a visible expansion of the particle size.
(3) The findings of the SEM demonstrate that the cobalt oxide nanoparticles have a shape that is semi-spherical.
(4) The resistivity of cobalt oxide nanoparticles reduces with an increase in temperature, indicating that these particles behave as semiconductors.
(5) Cobalt oxide, also known as Co_3O_4, was utilized to make the electrodes that were employed in the process of detecting hydrogen peroxide, also known as H_2O_2. The findings of the sensing demonstrate a great response toward the detection of hydrogen peroxide H_2O_2. Although there are a few different materials and methods that may detect hydrogen peroxide H_2O_2, the most successful method is to prepare cobalt oxide Co_3O_4 using a smooth hydrothermal process and then use it for hydrogen peroxide detection. This method will be described in more detail in future.

Acknowledgment

The authors would like to thank the Deanship of Scientific Research at the Umm Al-Qura University for supporting this work by Grant Code: (19-ENG-1-01-0013).

References

1. Z. Gong *et al.*, *Appl. Catal. B. Environ.* **305**, 121021 (2022).
2. J. A. Haider and N. Muhammad, *Int. J. Mod. Phys. B* 2250212 (2022).
3. J. A. Haider and N. Muhammad, *Int. J. Mod. Phys. B* 2250176 (2022).
4. S. Nadeem *et al.*, *Front. Phys.* 532 (2022).
5. R. Drasovean, S. Condurache-Bota and N. Tigau, *J. Sci. Arts.* **13**, 379 (2010).
6. T. Maruyama and S. Arai, *J. Electrochem. Soc.* **143**, 1383 (1996).
7. D. Cao *et al.*, *J. Power Sources* **179**, 87 (2008).
8. R. O'Hayre *et al.*, Fuel cell reaction kinetics, in *Fuel Cell Fundamentals* (Wiley, New York, USA, 2006), pp. 59–92.
9. M. Ando *et al.*, *J. Mater. Chem.* **7**, 1779 (1997).
10. Z. Zhang *et al.*, *J. Alloys Compd.* **392**, 317 (2005).
11. W.-Y. Li, L.-N. Xu and J. Chen, *Adv. Funct. Mater.* **15**, 851 (2005).
12. J. A. Haider and S. Ahmad, *Int. J. Mod. Phys. B* 2250217 (2022).
13. S. Asghar, J. A. Haider and N. Muhammad, *Int. J. Mod. Phys. B* **36**, 2250160 (2022).
14. L. Xie *et al.*, *J. Solid State Electrochem.* **17**, 55 (2013).
15. X. He *et al.*, *J. Mater. Chem. A* **1**, 1112111125 (2013).
16. X. Liu *et al.*, *Nanoscale* **5**, 6525 (2013).

17. G. Wang *et al.*, *J. Phys. Chem. C* **113**, 4357 (2009).
18. X. Xie and W. Shen, *Nanoscale* **1**, 50 (2009).
19. Y. Liu *et al.*, *Electrochim. Acta* **53**, 2507 (2008).
20. B. B. Lakshmi, P. K. Dorhout and C. R. Martin, *Chem. Mater.* **9**, 857 (1997).
21. R. Xu and H. C. Zeng, *J. Phys. Chem. B* **107**, 12643 (2003).
22. X. Shi *et al.*, *Nano Lett.* **2**, 289 (2002).
23. G. P. Anipsitakis, E. Stathatos and D. D. Dionysiou, *J. Phys. Chem. B* **109**, 13052 (2005).
24. J. Park, X. Shen and G. Wang, *Sens. Actuat. B. Chem.* **136**, 494 (2009).
25. M. Y. Nassar, *Mater. Lett.* **94**, 112 (2013).
26. W. Jia *et al.*, *J. Electroanal. Chem.* **625**, 27 (2009).
27. A. M. Amanulla *et al.*, *J. Photochem. Photobiol. B. Biol.* **183**, 233 (2018).
28. Y. Teng *et al.*, *Mater. Lett.* **64**, 239 (2010).
29. J. A. Dias, S. C. Maestrelli and M. R. Morelli, Evaluation of electrical, structural, thermal and optical properties of Co_3O_4 semiconductor *22^0 CBECiMat - Congresso Brasileiro de Engenharia e Ciência dos Materiais*, de Novembro de 6-10 (Natal, RN, Brasil, 2016).
30. S. A. Makhlouf *et al.*, *Superlattices Microstruct.* **64**, 107 (2013).
31. M. Salavati-Niasari, A. Khansari and F. Davar, *Inorgan. Chim. Acta* **362**, 4937 (2009).
32. S. Farhadi, J. Safabakhsh and P. Zaringhadam, *J. Nanostruct. Chem.* **3**, 69 (2013).
33. E. D. Lester *et al.*, *Prog. Cryst. Growth Charact. Mater.* **58**, 3 (2012).
34. D. A. Barlow, J. K. Baird and C.-H. Su, *J. Cryst. Growth* **264**, 417 (2004).
35. H. Itoh and T. Sugimoto, *J. Colloid Interface Sci.* **265**, 283 (2003).
36. M. Khalil *et al.*, *J. Nanoparticle Res.* **16**, 2362 (2014).

www.ingramcontent.com/pod-product-compliance
Lightning Source LLC
Chambersburg PA
CBHW071218060126
37764CB00004B/221
* 9 7 8 9 8 1 9 8 1 8 4 2 6 *